VK
597
C22
F5
KWANTLEN COLLEGE LIBRARY
S

The Chartmakers

This book is dedicated to the men and women of the
Canadian Hydrographic Service
To those who carry out the surveys
To those who produce the charts and other publications
To those who man the ships and launches
To those who provide engineering, technical and administrative support
And to those who keep the home fires burning.

The Chartmakers

The History of Nautical Surveying in Canada

STANLEY FILLMORE

R. W. SANDILANDS

Original Photography by Michael Foster

Published by NC Press Limited

in association with The Canadian Hydrographic Service

Department of Fisheries and Oceans

Published by NC Press Limited
31 Portland Street
Toronto, Ontario Canada M5V 2V9

Canadian Cataloguing in Publication Data
Fillmore, Stanley, 1925-
The chartmakers
ISBN 0-919601-92-8
1. Hydrographic surveying - Canada - History.
I. Sandilands, W. R., 1924- II. Title.
VK597.C22F5 1983 526.9'9 C83-098779-7

Printed in Canada

Design/Maher & Murtagh

Jacket Photography by Michael Foster.
(front)
C.S.S. *Baffin*, surveying uncharted waters of Labrador Narrows in the Canadian Arctic.
(back)
1. A hydrographic launch running a line of soundings. *(upper left)*
2. Planning a survey. *(lower left)*
3. A dory in Arctic waters — reminiscent of the first explorers' surveys. *(upper right)*
4. Sailing off Vancouver Island. *(lower right)*

Contents

Introduction

The history of any maritime nation is inextricably interwoven with the activities of the men who chart its coastal waters. Necessarily so. The country bordered by the sea may view its waters solely as a source of food and other resources; it may look upon the ocean as a trade route to and from other nations; it may regard the sea as both a source and a medium of supply. In any case, the presence of a sea coast forces a country to pose and answer certain questions: Are its coastal waters safe to sail, and if not, where not? In case of war can the shores be adequately protected? Where are the fish and a multitude of other natural resources most likely to be found?

The fate of a nation can turn on the answers to such questions. It is the responsibility of the men who chart the coasts — hydrographers — to supply the answers.

This book, *The Chartmakers*, traces the one hundred years of Canadian endeavour in charting the three oceans and the thousands of miles of inland waterways that provide this country with the world's most extensive network of navigable waters. It is a history of men, and of the instruments used in their work. But accompanying this major theme is a counterpoint that speaks to the very nature of the country itself and its people.

Almost three hundred years before the Canadian Hydrographic Service was born, Samuel de Champlain was charting the Saint Lawrence River seeking safe harbours for his shiploads of colonists. Through the first half of the eighteenth century French Jesuit missionaries taught hydrography to river pilots so that the rich harvest of Canadian furs could be safely shipped to Europe. It was James Cook and his fellow mariners of the Royal Navy who charted the same river and Cook's protegé, George Vancouver, who by charting the Pacific shores of Canada, brought the western limits of the country under the influence of the British crown.

Amongst their own kind, hydrographers are articulate; in public they are usually most reticent. As the text of this book explains, hydrography's greatest success lies in the lack of publicity — when no ships run aground, when newspaper headlines and television newscasts do not shriek of lives lost at sea.

On the occasion of its one hundredth anniversary, it is appropriate that the work of the Canadian Hydrographic Service — and of the men and women who have laboured in its name — be brought to a wider audience. It is my hope that the contribution of these thousands of devoted people will become more generally known through the publication of this book. Their contribution to the development and to the continuing history of Canada should be recognized and appreciated.

STEPHEN B. MACPHEE
Director General
Canadian Hydrographic Service
Ottawa, October 1983

The Legacy

Columbus found a world, and had no chart,
Save one that faith deciphered in the skies;
To trust the soul's invincible surmise
Was all his science and his only art.

George Santayana,
O World Thou Choosest Not,
1894

There is no land uninhabitable Nor any sea unnavigable.

Sir Hugh Willoughby
Sixteenth century British
sailor, soldier, explorer

A NICE BIT OF VERSE, PERHAPS. UNFORTUNATELY IT MISREPRESENTS THE facts of the matter. Christopher Columbus, on that first voyage in 1492, *did* carry with him some science and some scientific instruments — a compass, a quadrant, a traverse table. His "soul's invincible surmise" — an incorrect one: a belief that he was sailing to China — was not "his only art". And he did sail with a chart, or at least what in the fifteenth century passed for a chart. In fact, one of Columbus' "arts" was that of the chartmaker, an occupation at which he excelled both before and after his voyages of discovery to what eventually became known as the Americas.

Most of us are familiar with this idea of Columbus the great explorer, the discoverer of "the New World" (though his hold on that honour has been fairly well shaken of late). If he had also been represented to schoolchildren as a great chartmaker and a precursor of today's hydrographers, then the need for this book would not be so urgent. All would know what a hydrographer is, and the history and importance of hydrography in North America and Canada would be as familiar to the general public as that of its sister science, geography.

Hydrography, according to the *Encyclopedia Britannica*, is "the science dealing with all the waters of the earth's surface, including the descriptions of their physical features and conditions; the preparation of charts showing the positions of lakes, rivers, seas, the contours of the sea bottom, the position of shallows, deeps, reefs and the direction and volume, configuration, motion and condition of all waters of the earth."

To [the hydrographer] the interest is constantly kept up. Every day has its incidents. The accuracy of the work of each assistant, when proved, is an infinite gratification to him, and he has also the continual satisfaction of feeling that of all he does a permanent record will remain, in the chart which is to guide hundreds of his fellow-seamen on their way.

Rear-Admiral Sir William James Lloyd Wharton
Admiralty Hydrographer, 1884-1904

That definition covers a lot of territory. When one considers that two thirds of the Earth's surface is under water and that Canada itself has more coastline (fronting on three oceans) and more navigable rivers, lakes, and bays than any other country on the Earth, then the task of the hydrographer, and specifically of the men who have worked for the Canadian Hydrographic Service, appears to verge on the impossible. When one adds to the enormity of the task at hand the fact that Canada has been, and remains in many ways, dependent for its existence, its development, and the continued maintenance of its sovereignty particularly in the Arctic islands, on the work of its hydrographers, then a knowledge of the history of hydrography would seem to demand release from the world of the esoteric into the land of essential information.

This book, *The Chartmakers*, celebrates one hundred years of Canadian hydrography, from the time in 1883 when the Dominion government first accepted financial responsibility for a survey of Canadian waters, to the present day when the economic future of this nation remains, in many areas, dependent on the work of the Canadian Hydrographic Service. But in order to understand fully the achievement of those one hundred years it's necessary to examine and acknowledge the work of the men who first charted the waterways of Canada, men who, in the imperial service of Portugal, France, Spain, and Britain, laid the foundation for the building of a corps of Canadian hydrographers whose own standards of excellence will provide a similar legacy for the hydrographers of the next one hundred years.

The first European to sail in Canadian waters and to gather hydrographic information for the making of a chart — that is, a map of any waterway — was Giovanni Caboto, known more familiarly as John Cabot. Like Columbus, Cabot was an Italian navigator, sailing under patent of a foreign monarch, in his case, Henry VII of England. Leaving the port of Bristol in 1497, Cabot made his first North American landfall in June at either Newfoundland or Cape Breton. He, like Columbus, was searching for China and believed that his new-found islands lay just off the coast of Cathay.†

†Cathay was Marco Polo's name for part of what we know today as China.

Cabot returned to Canadian waters in 1498 and then, with his four ships and their crews, promptly disappeared, a fate not uncommon in the days of those early explorers who voyaged in strange waters in the small, clumsy, sailing ships of the day. Evidence that Cabot compiled some kind of chart, outlining the discoveries made on his first voyage, is contained in the earliest extant chart to show any portion of North America, the famous Juan de la Cosa Portolan World Chart, dated the year 1500. This portolan clearly shows the coasts, ill-defined to be sure, of Newfoundland and Labrador, and the waters around them are denoted as "the seas discovered by the English" (that is, John Cabot).

Portolans were the Renaissance equivalent of today's sophisticated charts. They were made by seamen navigators for the use of other mariners, and gained popularity during the thirteenth and fourteenth centuries as the Mediterranean navigators became accustomed to the magnetic compass, an invention imported from the Arab world. Prior to the introduction of the compass, European sailors had been reluctant to venture out to the open sea, and had restricted their voyages and their charts to the coasts of the Mediterranean and Atlantic around Europe and North Africa. The compass in the hands of the great Portuguese navigators such as Prince Henry and Vasco da Gama of the fifteenth century, opened the high seas to exploration. Its use resulted in the discovery and charting of the west coast of Africa, the subsequent passage around the Cape of Good Hope to India, and, of course, the discovery of the Americas and, eventually, the Pacific Ocean.

The portolan charts such as employed by Cabot and Columbus were characterized by their use of a system of straight lines emanating from the centres of compass roses scattered about the open spaces of the charts. These "rhumb" lines, denoting North, East, South, West, and the intercardinal points between, were included on the portolans (along with information on coasts, currents, harbours, shoals and winds) simply as lines of direction, or as primitive aids to navigation on the open sea. Omitted, however, from the portolan charts were the now familiar lines of latitude and longitude.

Latitude and longitude are to the modern hydrographer — to any map or chartmaker, for that matter — what language and syntax are to the poet, justice and laws to the jurist, anatomy and chemistry to the physician. They are the concepts, the ideas that govern the work of each. Without latitude and

The labours of the surveyor have always been and always must be the precursor of Commerce.

Rear-Admiral G.H. Richards
Admiralty Hydrographer, 1863-1874

longitude the mapmaker is literally lost. It is the job of the surveyor to be able to tell another navigator, by means of a chart, the exact location of a harbour, a reef, a shoal, a rock; and these locations are expressed in terms of the intersection of parallels of latitude, the imaginary horizontal lines which circle the globe, and meridians of longitude, the imaginary lines which run vertically over the surface of the Earth from pole to pole.

Contrary to popular myth, Columbus and Cabot were not unique in their belief that the world was *not* flat, that one would *not* necessarily sail west and drop off the edge of the Earth. As John Noble Wilford, in his book, *The Mapmakers*, notes: "The idea of a spherical Earth probably arose independently in many cultures, but as far as we know it was such influential Greek philosophers as Plato and Aristotle who established the idea in Western thinking. 'The sphericity of the earth', wrote Aristotle in *Meteorology*, in the latter half of the fourth century [B.C.], 'is proved by the evidence of our senses'." These evidences included the changes in the stars in the night sky as one travelled north or south, the disappearance of ships sailing over the horizon, and the curved shadow of the Earth on the moon during a lunar eclipse.

If we understand that to be a hydrographer one must first be a navigator (that is, be able to determine where one is on the face of the Earth, be able to determine where it is one would like to go, and then how to go about getting there), then we must understand also that to be a navigator — prior to the very recent introduction of electronic and satellite positioning systems — one had to have a good knowledge of astronomy, the first of the sciences. Aristotle's observations of the stars, moon and sun were not unique.

For at least five thousand years, beginning with the Babylonians, mankind had studied the heavens and plotted the sun and other stars in their courses. Through the centuries other ancient peoples added to the fund of knowledge, and in spite of false starts and notions that prevailed for a time but were later proven to be wrong, knowledge and science increased, accumulated.

By the second century A.D. western man was equipped with an atlas of the world and an instruction manual on the principles of astronomy, navigation, and cartography. Why then did we have to wait until the fifteenth century for the discovery of two continents and an ocean?

One major influence was the character of the Middle Ages, a time dominated by superstition and the suspicion that all science and scientific investigations were either heresy or, worse, witchcraft. The academic community at Alexandria, a leading scientific centre, was destroyed soon after the atlas was published. Noble Wilford notes that "in [A.D.] 391, Christians sacked the library, burned the priceless contents, and converted the shell into a church. It was a symbolic victory of faith over reason."

Ironically, however, it was Christianity which eventually re-opened the doors to exploration of the world. During the Crusades to the Holy Land of the Middle East in the twelfth and thirteenth centuries, the Europeans, from their contact with the Arabs, discovered the compass and the astrolabe, the latter being a device for measuring the altitude of the sun or a star, by which one can compute latitude. With these instruments the Arabs had been able to make crude charts of the coasts they visited on their trading voyages.

It was also during this time that the beginnings of the European Renaissance, that flowering of art and science, were fostered by the rediscovery of the writings of the ancient Greeks and Romans. The invention in Germany of printing with movable type allowed for the dissemination of these writings. By the fifteenth century the seeds of the science of hydrography had been carried away from their Mediterranean nursery and begun to blossom in those nations — Portugal, Spain, Britain and France — that were to dominate the exploration and charting of the world in general and Canada in particular.

The seaman's astrolabe was an early device for measuring the angular elevation of the sun or star above the horizon, a measure which permitted the sailor to calculate his latitude. The heavy brass circle was scribed in degrees, and the sighting rule pivoted about the centre. A ring at the top held the instrument vertical, either on the seaman's thumb or tied to a shroud.

In 1415 Prince Henry of Portugal, known as The Navigator, set up a naval establishment at Sagres and attracted the best navigators, mathematicians, astronomers, geographers, cartographers and instrument makers to instruct his captains and pilots. His pupils sailed off on voyages and brought back valuable cargoes, thus making exploration profitable. This in turn led to great activity in maritime commerce and a concomitant demand for better, more accurate, more informative charts. Publishers and princes gathered the descriptions; cartographers compiled the charts, maps and plans of the voyages and then disseminated them privately as rutters, or *routiers*, and eventually the portolan charts used by Columbus and Cabot.

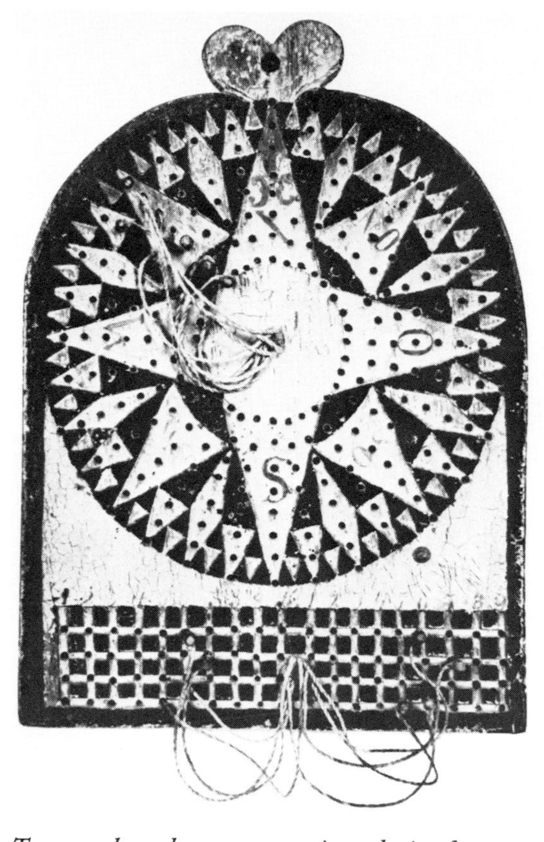

Traverse boards were an ancient device for recording the course run by a ship. A compass rose was painted on the board with a number of holes drilled into each compass point. Every half hour during each watch the helmsman inserted a peg into the hole that most closely approximated the course made good during the previous thirty minutes. At the end of each watch the mean course was calculated by "averaging" the accumulation of pegs.

Ultimately, the major impetus for voyaging west in search of Oriental spices and silks came from good old fashioned greed.

Indeed, the motivation for most hydrographic work over the centuries has been, and remains, the discovery and charting of safe and speedy sea routes both to the source of exploitable raw resources, and returning, to the marketplace where they may be traded for coin. For the Europeans in the late fifteenth century, matters came to a head when the Ottoman Turks completed their conquest of the eastern Mediterranean. This gave them control of territories through which ran the established overland caravan routes, and effectively cut off from Christian Europe the supply of spices, silks and other products of the Orient.

Voila! The Portuguese began to push further south along the west coast of Africa in pursuit of a sea route to India (Vasco da Gama arriving there in 1498) and the Spanish commissioned Columbus.

After word of his discoveries reached Europe, the Spaniards and their Portuguese rivals appealed to the Vatican to arbitrate a division of their anticipated holdings in the newly discovered territories. Pope Alexander VI, in 1493, decreed that all new lands east of a line drawn 370 leagues west of the Cape Verde Islands should belong to Portugal, and all those to the west of that line should be Spanish. This decision was ratified the following year as the Treaty of Tordesillas. Today, the most obvious result of the treaty is the fact that Portuguese is the language of Brazil, the most easterly of Latin American countries, while Spanish is spoken elsewhere in Central and South America.

Thus, on the 1502 Cantino Planisphere, the earliest map portraying any Portuguese claim to Canadian territory, we find that Newfoundland has been positioned well east of its actual location. In fact, it appears just on the Portuguese side of Pope Alexander's "line of demarcation", a handy, obviously politically expedient error.

During the next twenty to thirty years, the Portuguese continued to explore and chart the east coast of Canada. The new territories proved to be a lure not only for these skilled navigators, chartmakers and proto-hydrographers but also for fishermen who crossed the ocean in search of the great schools of cod flourishing in the waters of the Grand Banks. The Portuguese even made an attempt at colonization. Seymour Schwartz, in *The Mapping of North America*, notes that in 1520 Joaõ Alvares Fagundes, a Portuguese shipowner, "sailed along the south coast of Newfoundland and probably entered the Gulf of St. Lawrence". Five years later, Fagundes established a Portuguese settlement on Cape Breton. The fledgling colony was destroyed within the year by native Indians and by French fishermen. But Fagundes did leave his mark on the new land by discovering, charting, and subsequently bequeathing a bastardization of his name to the Bay of Fundy.

Published maps and charts of the New World changed little over this early period. On many, Labrador was still shown as connected to Cathay. On other maps the east coast of North America was seen as a continuous line from Florida to Newfoundland. The only significant advance appeared on a 1504 map by the Genoese, Nicolas de Caneiro, who was the first to replace the haphazard, compass rose lines of direction with a regular scale of parallels of latitude. And it was not until the voyages of the great French and British explorers of the sixteenth and seventeenth centuries that charts and maps of Canadian waters and lands began to bear any resemblance to the reality of the vast territories they claimed to represent.

Because of the 1494 Treaty of Tordesillas, exploration of the New World had been left mainly to the Spanish and Portuguese. However, the idea that a passage to China was available by sailing around or through the lands of the northwest – that is, Newfoundland, Labrador, and the Maritime provinces – persisted and attracted increasing attention from the governments of England, Holland, and especially France. They lobbied at the Vatican, and in due course Pope Clement VII declared that, in the eyes of God and the Church at least, the old line of demarcation applied only to those lands already discovered. It was as though a starter's pistol had been fired.

In April 1534, Jacques Cartier sailed from Saint Malo in France to Cape Bonavista in Newfoundland, over to the north entrance of the Gulf of Saint Lawrence, up the Strait of Belle Isle, back down to the gulf, to New Brunswick, and along the Gaspé Peninsula. There he encountered the Huron Indians and returned to France with two of the chief's sons.

On his second voyage, still in search of a passage to the East, Cartier sailed with three ships and one hundred and twelve men to the north shore of the

LES
VOYAGES
DE LA
NOVVELLE FRANCE
OCCIDENTALE, DICTE
CANADA,
FAITS PAR LE Sr DE CHAMPLAIN
Xainctongeois, Capitaine pour le Roy en la Marine du Ponant, & toutes les Descouuertes qu'il a faites en ce païs depuis l'an 1603. iusques en l'an 1629.

Où se voit comme ce pays a esté premierement descouuert par les François, sous l'authorité de nos Roys tres-Chrestiens, iusques au regne de sa Majesté à present regnante LOVIS XIII. Roy de France & de Nauarre.

Auec vn traitté des qualitez & conditions requises à vn bon & parfaict Nauigateur pour cognoistre la diuersité des Estimes qui se font en la Nauigation. Les Marques & enseignements que la prouidence de Dieu à mises dans les Mers pour redresser les Mariniers en leur routte, sans lesquelles ils tomberoient en de grands dangers, Et la maniere de bien dresser Cartes marines auec leurs Ports, Rades, Isles, Sondes, & autre chose necessaire à la Nauigation.

Ensemble vne Carte generalle de la description dudit pays faicte en son Meridien selon la declinaison de la guide Aymant, & vn Catechisme ou Instruction traduicte du François au langage des peuples Sauuages de quelque contrée, auec ce qui s'est passé en ladite Nouuelle France en l'année 1631.

A MONSEIGNEVR LE CARDINAL DVC DE RICHELIEV.

A PARIS.
Chez PIERRE LE-MVR, dans la grand' Salle du Palais.

M. DC. XXXII.
Auec Priuilege du Roy.

1632

Account of Champlain's voyages to New France between 1603 and 1629, published in 1632 under the auspices of Cardinal Richelieu.

Gulf of Saint Lawrence and then to the mouth of the Saint Lawrence River, where he met Indians who described the river as "chemyn de Canada", the road to Canada.† Cartier continued up the Saint Lawrence, accompanied by the chief of the Hurons, until he came to the village of Hochelaga where he named an overlooking hill Mont Réal. On this second voyage, Cartier's westward progress up the river was blocked by rapids which a later French explorer (Champlain), in a sarcastic jibe at Cartier's firm belief that China lay just the other side of this impediment, named Lachine.

On his third voyage, in 1541, Cartier tried to establish a permanent settlement at Cap Rouge, several miles upstream from Quebec City. This attempt at colonization failed within a year and the French, discouraged by this and Cartier's failure to find a passage to China, abandoned their territories in the New World to the fishermen of the east coast. All the while that Cartier had been exploring the Saint Lawrence, however, he had been charting the river and its coast and it was his hydrographic/cartographic work that eventually proved, according to Seymour Schwartz, "widely influential in contemporary mapping". The map of North America finally began to take shape later in the sixteenth century. Although the Europeans still had no inkling of the vast size of the lands of the northwest, they were beginning to suspect that if a passage to the Orient were to be had, it would more probably be gained by sailing over the top of the new territories, as opposed to attempting a route through the rivers of the continent.

Thus came the English under the banners of the East India Company and the Muscovy Trading Company and the patronage of Queen Elizabeth I. In 1576 Sir Martin Frobisher set off with three ships and a copy of the newly published Mercator Map of the World in search of the Northwest Passage. Due to errors on the Mercator chart, Frobisher, upon reaching Baffin Island, believed that territory to be Greenland, and Greenland he identified as one of the imaginary Friesland Islands. Because of this he claimed not Baffin Island, but the coast of Labrador for his Queen and named it West England. Hearing of this, the Portuguese, in a last ditch attempt at establishing some sort of

†Canada: "An Indian word that means a collection of houses, it first appeared on maps as reference to the village of Stadaconé at the present site of Quebec City.": Schwartz

sovereignty in North America, hastily made a title switch on their charts. They had previously claimed Greenland under the name of Lavrador, a Portuguese word meaning landowner, and simply transferred this appellation to the land west of Newfoundland and Baffin Island. Much to Frobisher's chagrin this bit of chicanery worked. Unfortunately for the Portuguese, only the name (corrupted in time, by the French, to Labrador) remained in place. However, Frobisher did sail as far as the southeast corner of Baffin Island and anchored in the bay there which still bears his name.

The next year, on his second voyage, Frobisher returned to Baffin, mined 200 tons of iron ore which he thought was gold, and carried it back with him to England. He made one more voyage but never passed west of Frobisher Bay.

The next English expeditions, in 1585, 1586, and 1587, were led by John Davis who travelled as far as Cumberland Sound and the Davis Strait. Henry Hudson made four voyages into the eastern Arctic between 1607 and 1611. On his last, his crew set him adrift in a lifeboat on Hudson Bay where, presumably, he died. William Baffin, in five expeditions between 1612 and 1616, penetrated to the northern end of the bay which bears his name. This period of exploration in search of a Northwest Passage ended with the voyages of Luke Fox and Thomas James, in 1631, by which time the British felt that it had been conclusively proven that there was no passage through the continent from the western shores of Hudson Bay.

In 1603 Samuel de Champlain made his first voyage to North America. This was when it all really began. This was the man who would at last bring some sense of order and system to the mapping of Canada's wilderness, the charting of her rivers and lakes. Like others before him, he always maintained, in the back of his mind, the possibility of finding a passage to the Orient. But as a future Lieutenant-Governor of New France — actually the first, appointed in 1633 — he was principally committed to surveying and charting in as much detail as possible the lands and waters under French control.

Champlain knew that to possess a territory effectively one must first know its geography. One must be able to state with authority, by putting it on paper

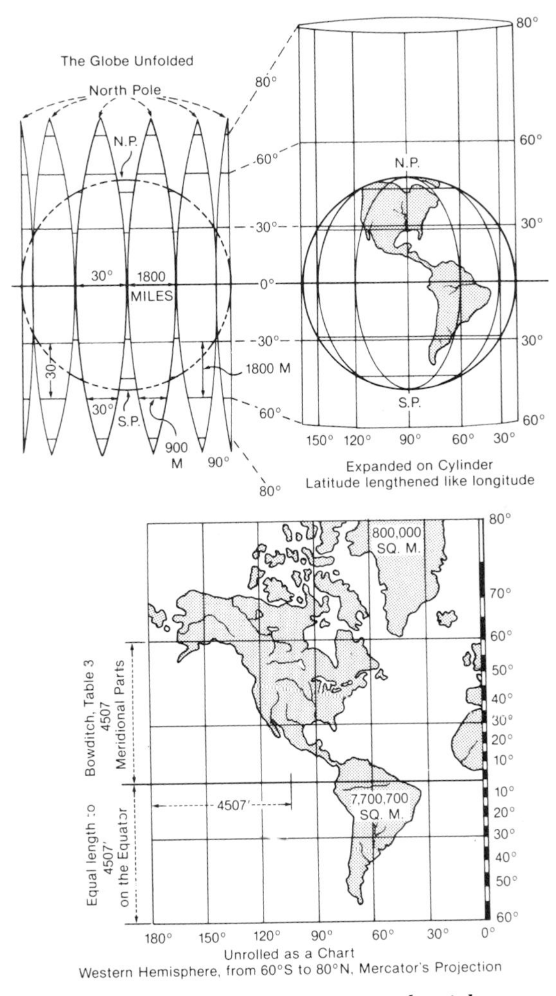

The genius of Gerardus Mercator, a Flemish cartographer of the sixteenth century, was in successfully "flattening" the spherical surface of the Earth so that it could be shown on a two-dimensional map or chart. His technique was to take gussets from the globe's surface and redraw the converging meridians of longitude as parallel lines. The Mercator projection, of course, distorts the apparent size of land masses, the distortion becoming greater as the distance from the equator increases. Still, Mercator's invention proved so valuable to travellers that most maps and charts even today still use it to show the Earth's surface.

in the form of a map or chart, that "this belongs to me". As an ardent advocate of colonization, Champlain understood that effective maps and charts were necessary aids to potential settlers. Ships bringing colonists needed safe, known harbours. Fur traders required a way of locating navigable rivers and lakes.

Although his surveys were primarily land-oriented, Champlain had seen service as a sailor and in his own writings observed at one point that "of all the most useful and excellent arts, that of navigation has always seemed to me to occupy the first place . . . this is the art which won my love in my early years."

For the next thirty years, until his death on Christmas Day in 1635, Champlain mapped New France from the tidal waters of the Gulf of Saint Lawrence to the shores of Georgian Bay. And though his main contribution was as a land surveyor, his chartmaking on the east coast of Canada and along the Saint Lawrence, from the mouth of the Saguenay River to Quebec City, set a standard of accuracy which far surpassed that of any work previously done in North American waters. Soon it would lead to the official establishment of French "l'hydrographie" in Canada.

As early as 1632, the French cartographers printed their first map of the Great Lakes, Champlain's "Mer Douce", and by 1650, Sanson, "Geographe du Roy" in Paris, had published maps of New France that showed the east coast from the Bay of Fundy, north to James, Hudson, and Baffin Bays. But these were maps rather than true charts and it was not until about 1678 that *cartes, avec sondes* (charts with soundings) were published for a part of the Newfoundland coast between latitudes 41° and 51° North.

Soon after Louis XIV assumed royal control of *Nouvelle France* in 1663, a program was instituted to produce a body of ships' masters and pilots in the colony. The development of this plan had two aims: instruction in hydrography to produce surveyors capable of making accurate charts, and secondly, the training of masters and pilots capable of using these charts to navigate safely up and down the Saint Lawrence.

In 1681, Colbert, at that time the French minister of marine, decreed that all river pilots and sea captains were to observe and report on soundings and other hydrography in the Maritimes and Quebec. Four years later the first

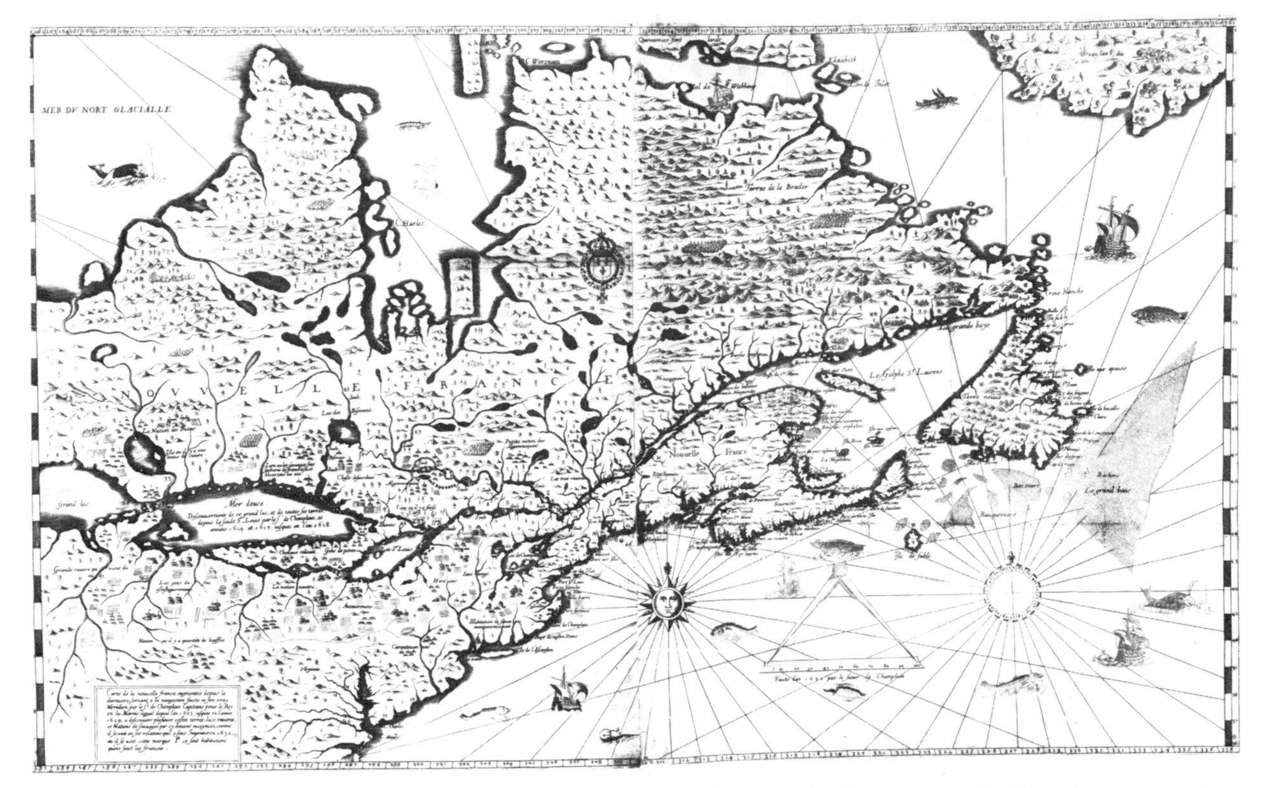

Champlain's map of eastern Canada, while inaccurate by modern standards, is a remarkable achievement for the instruments and techniques he used at the turn of the seventeenth century.

Canadian-based "Hydrographe du Roy", Sieur J.B. Franquelin, was appointed by the king.

Franquelin had been trained in cartography at the Department of Marine in Paris and was undoubtedly the best trained mapmaker in New France. But he was not an outstanding teacher and as time went on it became obvious that he was not comfortable in this role. His situation was probably exacerbated by the fact that he was not paid for his teaching duties until twelve years after receiving the appointment. The Governor of New France, the Marquis de Denonville, who had initially recommended Franquelin, wrote finally to the French marine minister suggesting that Franquelin be relieved of his teaching commitments and be employed exclusively in cartography and hydrography.

It was thought that perhaps the training of hydrographers and pilots could be better handled by the Jesuits; however, at that time (1687) nothing came of the suggestion.

With the return to New France in 1689 of Louis de Buade, Count Frontenac, for his second term as governor, Franquelin expressed again his desire to be rid of pedagogical duties and the shrewd Frontenac had him recalled to Paris where he was given an appointment as a cartographer. The Quebec post was thus left open for his successor, Louis Jolliet, a native-born Canadian who had been a contender for the post in 1685. However, upon Jolliet's sudden death in 1700, the dispatch ships carried requests and recommendations for a new appointee. To the chagrin of the Canadian authorities came the reply that Franquelin had been reappointed. But he never did return to Canada and instead, Jean Deshayes, perhaps the finest hydrographer in Quebec of the late seventeenth and early eighteenth centuries was given the appointment in 1702.

In the meantime, the Jesuits had begun giving instruction in navigation and pilotage, and had requested that they be officially commissioned to carry out this duty. Indeed, during the short time that the native Quebecois, Jolliet, held the job of royal hydrographer, he was associated in a minor, lay capacity with the Society of Jesus. However, it was not until the priests gave the government assurances that some young seminarian, rather than an elderly missionary, would lead the instructors that they were given this responsibility in 1706. They held the appointment for their college at Quebec City until the end of the French regime in Canada. These Jesuit instructors were supposed to have demonstrated their theoretical skills by practical application, carrying out field surveys on the Saint Lawrence River. But the teaching demands on their time were such that they produced little in the way of concrete results, in the form of completed charts. The surveys of Jean Deshayes, on the other hand, were eventually compiled for the production of the first published chart of the Saint Lawrence, a work which, according to naval historian James Pritchard, "stands as a precursor of the modern hydrographic chart".

Deshayes had been appointed an "Engineer to his Majesty for Hydrography" in 1681 and was a mathematician and surveyor rather than a ship's

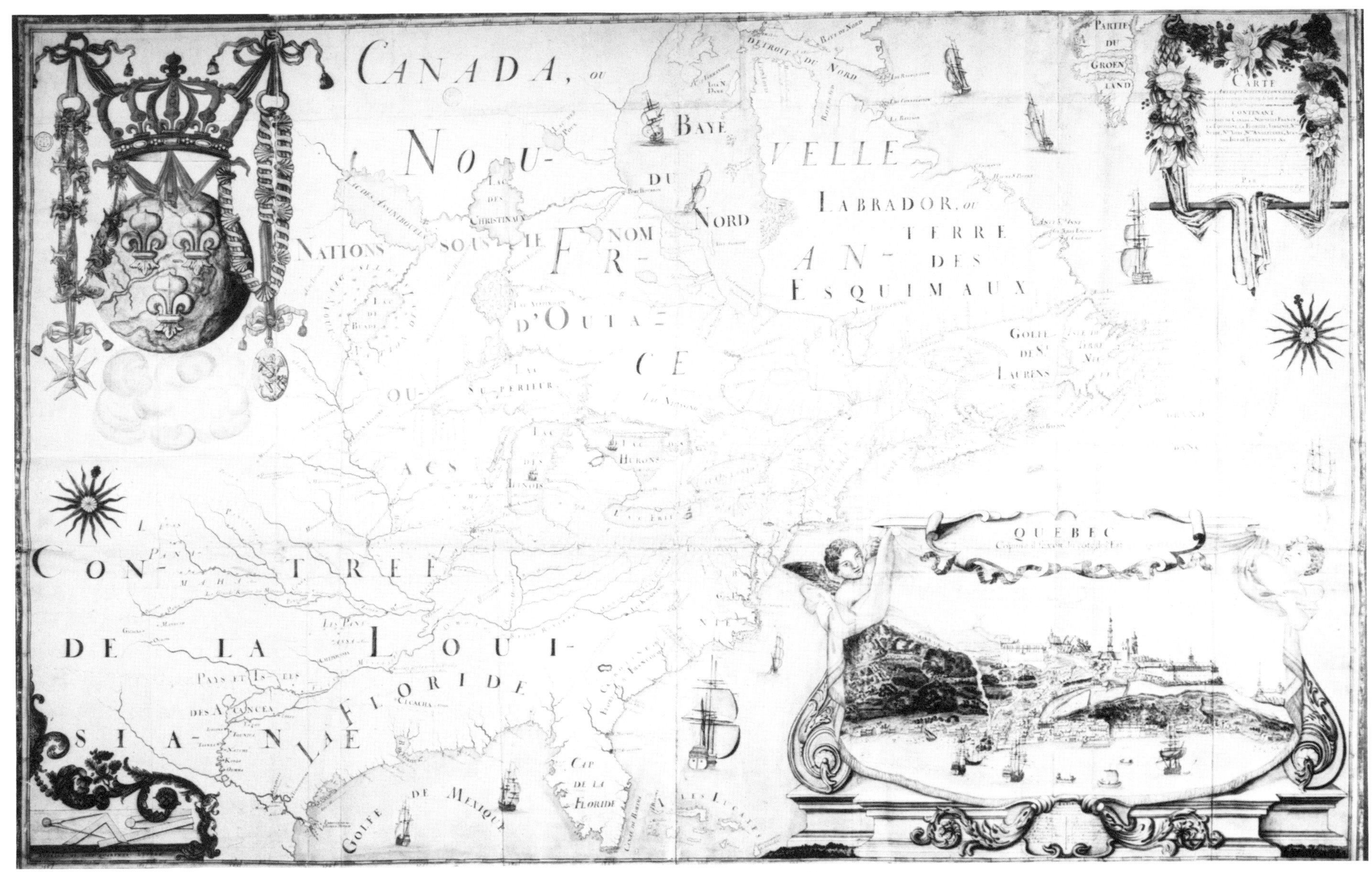

The first of the French "royal hydrographers" to be stationed in New France, Sieur J.B. Franquelin was a supremely gifted cartographer who produced this detailed map of North America in 1688. Though distorted, particularly in its horizontal dimensions, it represents a high point of the mapmaker's art.

master or pilot. He had been active and was well known in the French scientific community before arriving in Canada in 1685 with his commission as royal hydrographer. His survey of the Saint Lawrence River occupied him from May to November 1686, and was incomplete when he sailed for France with the intention of returning to Quebec the following year to finish the job. But he was not to see Canada again until 1702 when he was appointed professor of hydrography at Quebec.

What set Deshayes apart from previous French hydrographers were his attention to detail and accuracy, his use of the most modern instruments and methods available, and his devotion to the methodical. He started his survey on foot (on showshoes) during the winter and sketched the coastline of the south shore of the Saint Lawrence as far downstream as Rivière Quelle, seventy miles from Quebec City, and to Cape Tourmente on the north shore. He oriented all his sketches by means of a box compass and recorded his paces to provide a check on his estimated distances. Later on, during the summer, he sketched from a canoe offshore, again estimating his distances and landing for periodic checks where dangers to navigation required closer measurement.

Deshayes was hindered by having no qualified assistants, and in places where he did leave a boat crew to take soundings while he went ahead with his sketching, he sometimes found that he had to return and resound several areas because the hired sailors often invented the figures rather than risk the danger of sounding in shoal areas. Eventually this led Deshayes to chart the crews' soundings in Roman numerals, in order to distinguish their figures from his own more accurate and reliable ones.† Though he took many soundings, his chart ultimately showed only the dangerous areas, the contours of the sandbanks and a few depths in mid-channel. His preliminary copy of the chart was

†History, it has been noted, repeats itself and more than two hundred years later the British Admiralty adopted a similar technique to distinguish between leadline soundings and those obtained from the newly introduced echo sounders. The new device, in the eyes of the Lords of the Admiralty, did not produce the same precision as the leadline. In time, however, the echo sounder proved itself and the distinction on British charts between the two kinds of soundings was eliminated.

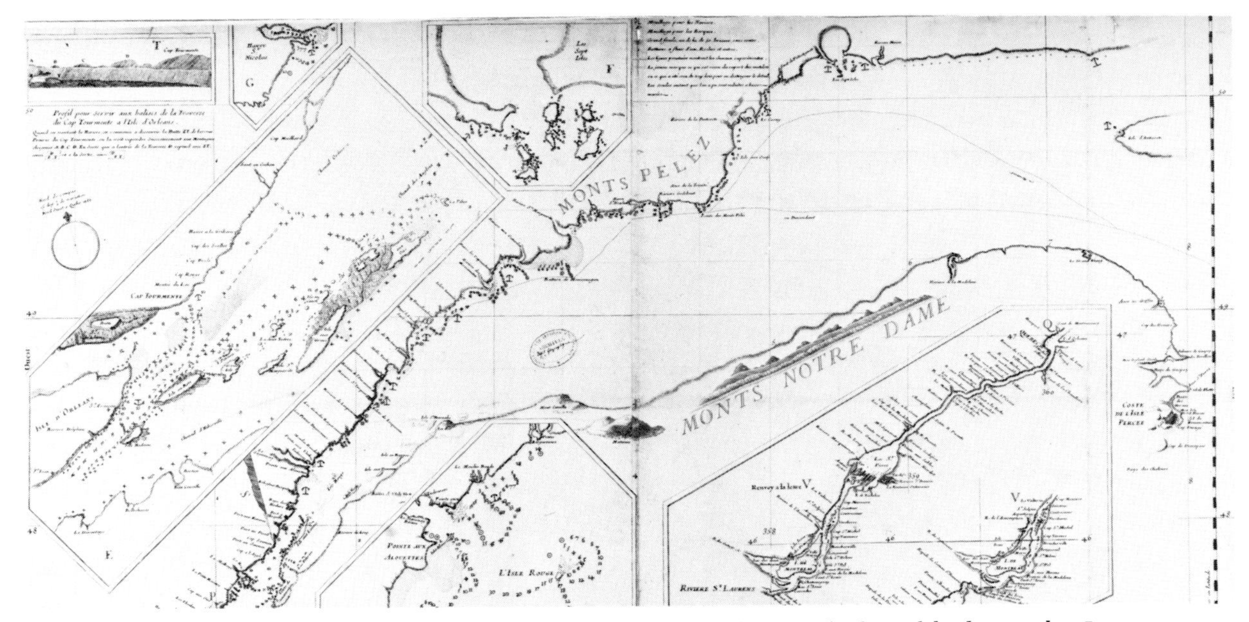

Canadian hydrography achieved the status of a profession with the arrival of royal hydrographer Jean Deshayes in 1685. He was the first to apply the rigourous standards of accuracy that chartmaking requires, and he made full use of the range of instruments available to him. His chart of the Saint Lawrence from Quebec City to the gulf has been hailed as the "precursor of the modern hydrographic chart."

about five feet long, included two detailed insets, and was accompanied by written "sailing directions", including compass variations and tables giving the hours of high tide and the days of the full moon, for twenty-four locations on the north shore, and other advice for the mariner.

Jean Deshayes died at Quebec in 1706 and, though he worked in Canada for only four years, his chart became the standard along the Saint Lawrence for the rest of the French period. In fact, in 1757, when the British produced their own first preliminary chart of the river, it was based on Deshayes's survey.

In 1714, using the ship *Afriquain*, Sieur de Voutrain, another French hydrographer, made a running survey of the Saint Lawrence from Isle d'Orléans to Kâmouraska. Also, about that time, a plan was approved by the Duc d'Orléans which proposed having a frigate and two small boats operate on the river for two years for hydrographic purposes. But the financial drain of building and maintaining the fortress at Louisbourg effectively scuttled the scheme.

In the profoundest sense, Captain James Cook, RN, shown here in his official portrait by Sir Nathaniel Dance, established the guidelines by which hydrography in Canada and throughout the world progressed from the status of an art to science. With his work in the south Pacific, and on both east and west coasts of Canada, Cook codified and standardized the rigid practices by which future hydrographers were to work.

Then in 1720, the French Hydrographic Office or "Depot des cartes, plans et journeaux", was established in Paris. This action indicated an awareness on the part of the French of the importance of the collecting and publishing of surveys and charts for the safe navigation of their naval and mercantile fleets. Plans were made to survey the Gulf of Saint Lawrence and the Canadian coastal waters on the Atlantic to aid the development of fisheries and trading posts. Again little came of these schemes.

However, the loss of the transport ship *Eléphant* on the shoals off Cap Brulé in 1729 brought matters to a head and a new program of surveying was initiated by the French. This initiative had really already begun in 1727 with the appointment of Richard Testu de la Richardière as port captain at Quebec. He was given the added duty of being responsible for navigation on the whole of the Saint Lawrence. Accordingly, each year, one or two of the navigators who brought in the annual supply convoy from France were left at Quebec to assist in surveys and thus better acquaint themselves with the navigation of the river. These surveys and their resultant charts were never officially published, but numerous copies were made for use by the French men-of-war. La Richardière also carried out surveys in the Strait of Belle Isle, around Isle Saint Jean (Prince Edward Island), Baie des Chaleurs, and the Strait of Canso. By the time of la Richardière's death in 1741, the charting was adequate for the French colonial trade.

Had an able successor to la Richardière been appointed immediately after his death, and the surveying impetus continued, better French charts might have become available. But the position of port captain was allowed to lapse for a few years, and of the final three appointees to the post, the first died shortly after his arrival, the second never put in an appearance at Quebec, and the third viewed the position as a well-earned retirement perquisite. In 1751, during the term of this last official, an able man called Gabriel Pellegrin, who had worked with la Richardière, was appointed assistant port officer. But his efforts came to little in the face of the lethargy and jealousy of his superior.

Finally, in 1759, during the closing months of the French regime in Quebec, Pellegrin assisted in the development of defence plans for the Saint Lawrence River. However, the French had no real intention of using the navy in the coming battles and Pellegrin's suggestions were never implemented. His last desperate attempts to contribute something to the war effort were directed at removing navigational buoys in the "Traverse" on the Saint Lawrence where the English ships were expected to pass, replacing them with false aids to navigation. As we shall see, this trick did not work.

Shakespeare wrote that "There is a tide in the affairs of man, which taken at the flood, leads on to fortune." The tide of fate brought two men to Louisbourg in 1758. They were part of the British forces gathered to wrest this principal fortress from the French.

Samuel Holland was an army engineer under the command of General James Wolfe and was well qualified to prepare maps of the area.

The other man was the sailing master of HMS *Pembroke*, one of the blockading ships. He was to take his place in history as the father of modern hydrography.

James Cook was an anomaly in his time. The son of a lowland Scots labourer and a Yorkshire village woman, he was born in Marton-in-Cleveland in the North Riding of Yorkshire in 1728. He received the rudiments of his education in return for services as a farmhand and eventually became shopboy to a grocer and haberdasher. A life of commerce was not for Cook, however, and he moved to Whitby and apprenticed himself to a shipowner and coal exporter.

In this period there were more than a thousand coal carriers engaged in shipping Tyneside coal, and the trade served as a nursery for British seamen. It was a hard nursery, but it graduated sailors with an expert knowledge of ships and the sea. The more apt pupils learned of the hazards of navigating the treacherous east coast of England which was unlit, unbuoyed, and only roughly charted. They learned to "read" the water for the shifting shoals of the Thames Estuary and the offshore dangers of the banks and rocks of the North Sea with its storms, fogs and unpredictable tides and currents.

On this coasting trade and its North Sea crossings, Cook learned the four basic *L*'s of the science of navigation: lookout; lead (the weight on the end of a sounding line); log†; and latitude. These principles stood him in good stead for his three great voyages of discovery later in life.

His apprenticeship over, he worked his way up to employment as mate and was offered his own command, a position which promised a comfortable income and future. Unpredictably, he turned down this offer and volunteered for the Royal Navy at Wapping on 17 June 1755. Life in the British navy of the eighteenth century was not easy. Admiral Edward Vernon, writing of the time, opined that the fleet was "manned by violence and maintained by cruelty". Ships were often crewed by the press gangs who virtually kidnapped young men since few trained seamen volunteered. The opportunities for promotion from the lower deck were negligible unless one had a patron and Cook's relatives and friends included no earl or duke who had the ear of the First Lord of the Admiralty.

Cook's abilities were soon recognized, however, and within a month he was rated master's mate. The one advantage Cook found in the Royal Navy was the opportunity provided there for formal tutoring in the arts of navigation. During the reign of Queen Anne, official regulations had been laid down by the Admiralty for the schooling in navigational theory of the "young gentlemen volunteers" or midshipmen aboard Royal Navy vessels. Cook thus began studying for his master's warrant while at sea with the navy. It was in the course of these studies that he added a knowledge of the rudiments of hydrography to his previously gained store of navigational skills. In the remarkably short time of two years, Cook was awarded his warrant from the naval examiners of Trinity House. He was then appointed Master of a Royal Navy ship, the 64-gun, 1250-ton, HMS *Pembroke*.

The position of master of one of his Majesty's ships was that of the chief professional sailor, responsible to the captain for the ship's navigation, gen-

†Log is the name given to any device used to measure a ship's speed through the water, or for measuring the distance travelled in a given time; a ship's log is thus comparable to a car's speedometer/odometer. In the beginning the log had been precisely that — a log or a block of wood to which was attached a considerable length of line. When the log was thrown overboard from the ship's stern, it was assumed to remain where it landed; by measuring the length of the line run out in a given time, the speed of the ship could be easily calculated. To facilitate the calculation the line was knotted at regular intervals, and the knots counted, rather than a measuring rule applied to the line. Thus arose the term *knot* referring to a ship's speed.

eral administration, the taking of soundings and bearings which corrected or added to the charts available, and the collection of materials for publication in *Sailing Directions*, the printed instructions which elaborate on the charts' information.

The value of a good master was incalculable and for this reason he tended to remain a master. It has been said that Cook was denied a commission for many years because of his common background. But it was probably mainly because of his high professional capabilities. Why lose a first class master by giving him a commission and thus promote him out of his area of proven competence?

There is no doubt of Cook's abilities. Still, there is an element of luck in the life of every great man. Perhaps he has a fortuitous meeting, is at the right place at the right time and becomes great because he has prepared himself for opportunity and knows how to turn it to his advantage when it presents itself.

Cook was fortunate in his appointment to HMS *Pembroke*, as she was commanded by Captain John Simcoe, one of the most intellectual officers in the navy, a man of scientific leanings who took a small but select library of navigational books to sea with him.

Louisbourg surrendered on 26 July 1758 and the fateful meeting between Cook and Holland took place the following day at Kennington Cove.

Cook had gone ashore and his curiosity was aroused by an officer carrying a small square table mounted on a tripod; he would set this down, sight along the top in many directions and then write in a notebook. The two men struck up a conversation and Cook discovered his chance companion was Samuel Holland, engaged in making a Plane Table survey of the encampments. Cook expressed a desire to learn more about surveying and in the words of Holland:

> I appointed the next day in order to make him [Cook] acquainted with the whole process; he accordingly attended, with a particular message from Captain Simcoe, expressive of a wish to have been present at our proceedings; and his inability, owing to indisposition, of leaving his ship; at the same time requesting me to dine with him onboard; and begging me to bring the Plane Table pieces along. I with much pleasure accepted that invitation, which gave rise to my acquaintance with a truly scientific gentleman

> [Simcoe] for the which I ever hold myself much indebted to Captain Cook. I remained that night onboard, in the morning landed to continue my survey at White Point (the other end of Gabarus Bay), attended by Captain Cook and two young gentlemen.

The initial tuition may have lasted longer than one day, as the *Pembroke* was moored in Louisbourg for most of August before departing to the Bay of Gaspé and the Gulf of Saint Lawrence on raids on the French settlements in these areas. But Cook immediately applied his newly acquired knowledge, and his first engraved and printed chart was of Gaspé Bay and Harbour, dedicated "by James Cook, Master of His Majesty's Ship the Pembroke" to the master and wardens of the Trinity House of Deptford.

With the onset of winter the fleet retired to Halifax and whenever Holland could spare himself from his duties he went onboard *Pembroke* where Simcoe, Cook and he compiled a chart of the Gulf and River Saint Lawrence. A second chart of the river, including Chaleur and Gaspé Bays, was also compiled and sent to England for publication.

Again we quote Holland writing to Captain Simcoe's son, John Graves Simcoe, who became the first Lieutenant-Governor of Upper Canada:

> These charts were of much use, as some copies came out prior to our sailing from Halifax for Quebec in 1759. By the drawing of these plans under so able an instructor, Mr. Cook could not fail to improve and thoroughly brought in his hand as well in drawing as in protracting, etc., and by your father's finding the latitudes and longitudes along the Coast of America, principally Newfoundland and Gulf of St. Lawrence, so erroneously heretofore laid down, he was convinced of the propriety of making accurate surveys of those parts. In consequence, he told Captain Cook that as he had mentioned to several of his friends in power, the necessity of having surveys of these parts and astronomical observations made as soon as peace was restored, he would recommend him to make himself competent to the business by learning Spherical Trigonometry, with the practical part of Astronomy, at the same time giving him Leadbitter's works, a great authority on astronomy, etc., at that period, of which Mr. Cook assisted by his explanations of difficult passages, made infinite use, and fulfilled the expectations entertained by your father, in his survey of Newfoundland. . .

But first, Cook's services were required for an expedition that involved navigating some 400 miles up the Saint Lawrence with charts that were inadequate to the needs of the size and numbers of ships required to defeat Montcalm and capture Quebec.

Though some French charts had been captured, no great stock was placed on their accuracy. So the prisons of Britain were scoured for French Canadian prisoners-of-war who had knowledge of the river pilotage. These men were forced into service along with at least seventeen other pilots familiar with the Gulf and River Saint Lawrence from Louisbourg to Gaspé, Mont Louis and Grand Rivière.

The chart compiled by Holland and Cook that winter would today be classified as provisional, but it sufficed. No problems were anticipated through Cabot Strait and the Gulf of Saint Lawrence as far as the present-day Pointe-au-Père. From there to Tadoussac there was deep water as long as one avoided the shore on either hand. But then the dangers multiplied and after passing the mouth of the Saguenay, the hazards of shoals, drying banks and bewildering currents grew. The channel ran inside Ile aux Coudres, close to the northern mainland shore, though the greater width of the river lay south of the island. It then crossed diagonally to the northeastern end of Ile d'Orléans and finally passed between the eastern shore of that island and Little Ile Madame and round the bare rock of Orléans into the basin of Quebec.

It was this diagonal crossing, known as the Traverse, that was the greatest hazard. It was inadequately charted and the French had never brought a large ship through. The French pilots knew its dangers and it was crudely buoyed and marked but these aids to navigation had been removed and altered by Pellegrin in the face of the anticipated British assault.

Admiral Durrell, commander of the fleet, left Halifax on 5 May 1759 and worked his way through the ice and up the Saint Lawrence under a favourable wind. He left a few of his ships at Bic but took the majority as far as Ile aux Coudres. Four vessels, one of which was the *Pembroke*, were sent further ahead to Ile d'Orléans and by 8 June they were at the downriver end of the Traverse. For the next two days all boats were out sounding the passage. History has credited Cook with being the main participant in this operation but all the masters assisted in the survey.

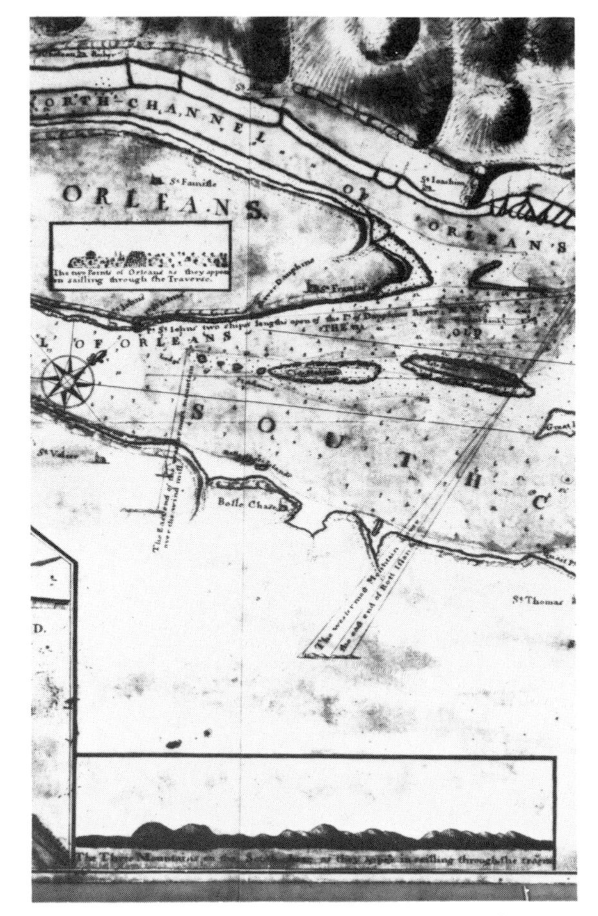

Cook's chart of the Saint Lawrence at Quebec City, particularly of the treacherous Traverse, permitted Wolfe to bring his troops upriver and successfully wage the Battle of the Plains of Abraham. Cook is generally credited with production of this chart survey but, in fact, all the sailing masters in the fleet contributed.

While these preliminaries were afoot, Admiral Charles Saunders, who was in overall command of the naval forces, with its great fleet, was slowly moving up the Saint Lawrence, and by 18 June his nine ships of the line, thirteen frigates and one hundred and nineteen transports had reached Bic. One week later, they passed through the Traverse. Cook's log for 25 June records "at 11 a.m., a signal for all boats, manned and armed, in order to go and lay in the Traverse as buoys for the ships to come up."

The navy had safely brought General Wolfe and his troops before the fortress of Quebec.

Now the landing beaches had to be surveyed. This work frequently took place under fire. Cook himself narrowly escaped injury while making beach reconnaissances when a party of French and Indian canoes cut off his sounding boat.

Wolfe made his assault on 13 September 1759 and the battle of the Plains of Abraham sealed the fate of the French power in Canada.

Later that month Cook transferred to HMS *Northumberland*, and the following months found him again wintering in Halifax, whiling away the winter hours, furthering his studies of mathematics and astronomy. He also spent time drawing charts and writing *Sailing Directions*; it is to this period that his three manuscript charts of the area still extant must belong.

The next summer found Cook again at Quebec and the arrival of the relief ships broke the French siege. Lieutenant-Colonel Jeffery Amherst's advance from New York up the Mohawk River, across Lake Ontario and down the Saint Lawrence brought about the surrender of Montreal on 7 September 1760.

Back again in Halifax for the winter of 1760-1, an entry in the commodore's journal for 17 January notes that the master of the *Northumberland* was paid fifty pounds for his indefatigable industry in making himself master of the pilotage of the River Saint Lawrence. It was a handsome bonus when one realizes that a master of this class of vessel had a regular pay of about six pounds per month. It was also an indication that Cook was attracting the attention of his senior officers.

In a last flurry of action in North America the French seized Saint John's, Newfoundland, in July 1762, and a British squadron was dispatched to recapture the port. This was quickly accomplished but its importance as it concerns Cook was that once again he met the right man at the right time in his career. The man was Joseph Des Barres, another military surveyor on Amherst's staff. Cook worked with him, learned more of surveying, and together they charted Conception Bay where the British fisheries were to be re-established and extended.

Des Barres, of a noted Huguenot family, was a Swiss who had studied mathematics and later enrolled as a cadet at the Royal Military College at Woolwich, England, where he applied his mathematical knowledge to surveying and fortification engineering. In this case the transfer of knowledge was two-way, for while Cook learned more of land survey control, Des Barres learned of hydrography and within the next few years himself became a competent hydrographer working under the general direction of the Admiralty, the British Hydrographic Office not yet being established. Between 1777 and 1784 he compiled and issued as a set of charts, *The Atlantic Neptune*, comprising the surveys of the eastern seaboard of North America.

In 1763, when the Treaty of Paris, ending the Seven Year's War, was signed, France lost most of her North American colonies but was allowed to re-occupy the offshore islands of Saint Pierre and Miquelon.

Later that same year, at the request of Captain Graves, Cook was selected to conduct hydrographic surveys of Newfoundland. He was appointed "King's Surveyor" and with a small vessel, the *Grenville*, a 68-ton, Massachusetts-built schooner with a crew of seven, he spent the next four years surveying on the Newfoundland coast where he produced his finest hydrographic surveys.

Cook invented nothing and he originated nothing, but he came on the scene at the time of a flowering of scientific knowledge and instrument making. He applied himself to achieve a complete understanding of the latest methods of navigation, including the use of the sextant and Harrison's newly invented ship's chronometer. He made sure that he was accompanied by trained astronomers on his three voyages in the Pacific and he learned from them all he could. Cook had one great advantage over those who took their master's papers at

The persistence of the Gallic presence

The British victory at Quebec did not terminate the French presence in Canadian waters. The Treaty of Utrecht in 1713 and the Peace of Versailles (1783) gave the French fishing rights along a stretch of Newfoundland's northern and western coastline which came to be known as the "French Shore". After the Plains of Abraham the French fishing fleet continued to exercise these rights. But by 1825, some sixty years and more after Quebec, the French government dispatched a survey ship to survey harbours along the French Shore. The surveys continued for about forty years. The approximately one hundred and fifty charts produced were for the use of French fishermen. But they were seen, by officials at the British colonial office at least, as an assertion of French sovereignty over part of Newfoundland. The French Shore became a source of friction between Britain, France and Newfoundland. It wasn't settled until 1904 when France renounced her rights under the Treaty of Utrecht in exchange for colonial concessions in Africa. Evidence of these early French surveys still show on some CHS charts of Newfoundland. Britain used some of the French data to improve Admiralty charts. After Newfoundland joined Canada, most of the British charts were turned over to Canada and reproduced here. Some CHS charts of Newfoundland are still based on the French surveys.

the same time as he, in that he was able to draw upon his years of experience in the practical application of navigational theory while he had been in the merchant marine.

Cook's contribution to the art of hydrography is stated simply: by the application of rigid standards of accuracy he converted hydrography from a craft to a science.

For example: determining a ship's exact position, in the face of the difficulties of merely steering the vessel properly, was regarded as a necessary but essentially secondary function by most masters. In Cook's day there were a number of methods and a variety of instruments to determine the accurate location of the ship. But most sailors would choose the simplest and easiest way. It was not that they were necessarily lazy but, they argued, the uncertainties of navigation were so many – leeway, currents, the difficulties in calculating longitude – that to attempt high precision was a waste of time.

This was *not* Cook's attitude. He believed that one of his duties as an explorer/navigator was to produce accurate charts of the waters he sailed. As a hydrographer he took great pains to establish his position accurately in terms of latitude and longitude.

To take one example: the standard method of surveying a coast, prior to Cook's innovations, had been the running survey. As the ship sailed along the coast her course was logged and plotted. The outstanding features of the coast were then positioned by cross bearings taken from the ship, and the shape of the coastline between these conspicuous landmarks was filled in with careful sketching. Soundings were taken from the ship, frequently of the "no bottom at — fathoms" variety. When time allowed, boats were lowered, additional soundings were taken, and greater detail of the coastline noted. At best, however, these were reconnaissance surveys, and there were inherent errors due to the difficulties of accurately fixing the ship's position and determining her speed.

Cook, with his recently gained knowledge of cartography and land surveying, changed these methods for his survey of Newfoundland.

He went ashore and made his observations from the stable platform of land instead of the rolling deck. He accurately measured his survey base lines, and extended a network of "triangulation" in the same way as a land surveyor would. He climbed hills to attain a better appreciation of a sinuous coastline. What better masthead than a high hill; and few sailors could have climbed as many hills as Cook. Using his precisely observed points, Cook sketched in the coastline. Then, with the observed points and an accurate outline of the coast, he positioned his ship and her boats and only then began to sound the waters.

The schooner *Grenville* moved from harbour to harbour in Newfoundland, the boats sounding and Cook, with his instruments onshore as much as possible, measuring, observing, remeasuring and positioning his flagged markers to be used as beacons and positioning aids in the sounding operations.

Admiral Sir William James Lloyd Wharton, Hydrographer of the Royal Navy from 1884 to 1904, later wrote of Cook's Newfoundland surveys:

> The charts he made during these years on the schooner *Grenville* were admirable. The best proof of their excellence is that they are not yet superceded by the more detailed surveys of modern times. Like all surveys of a practically unknown shore and especially when that shore abounds in rocks and shoals, and is much indented, they are imperfect in the sense of having many omissions. But when the amount of ground covered, and the impediment of fogs and bad weather on that coast is considered, and that Cook had at the most only two assistants, their accuracy is truly astonishing.

The surveys Cook carried out in Newfoundland were undoubtedly his finest "pure" surveys. He had the time and the equipment to give of his best. His other surveys throughout the world were masterful examples of the running survey but the goals set for him on these expeditions were different. On his voyage to the west coast of Canada he was sent primarily as an explorer in search of the elusive yet still alluring Northwest Passage.

> I rejoice . . . that we men of insular origin are about to recover one of our lost senses — the sense that comprehends the sea.
>
> Thomas D'Arcy McGee, 1864

In 1778 under westerlies punctuated by gales from the south with rain and sleet his two ships closed the land. A vista of heavily forested hills and mountains capped with snow presented itself. Such was Cook's first view of Vancouver Island, off the Canadian west coast, so different from the Maritimes and Newfoundland on the other side of the continent, where he had learned his surveying and made his initial impression on the naval authorities of the day.

After a five-week crossing of the Pacific where scarcely even a seabird was seen, Cook was anxious to take on fresh water and make repairs to the *Resolution's* foremast.

He had sailed many miles and achieved an international reputation since his days on Canada's east coast. On his first great voyage (1768-71) he sailed the ship *Endeavour* round the world, west about, with the main objective of observing the transit of Venus from the recently discovered Pacific island of Tahiti. Leaving Tahiti, he had sailed southwards in the search for a great southern continent. He reached latitude 60° South without any sign of this supposed continent and circumnavigated New Zealand, then followed the east coast of Australia, crossed the Indian Ocean and, after calling at Cape Town, South Africa, returned home to England.

On his second voyage (1772-75) with the *Resolution*, a 100-foot, refitted coal carrier, Cook sailed east to west calling at Kerguelen Island in the south Indian Ocean and on to New Zealand and Tahiti. It was on this voyage that he ventured further south than any man had before this time, reaching beyond the 71st parallel. At the point where he turned north again, young midshipman George Vancouver ran to the end of the bowsprit and exclaimed with a gleeful shout of *"ne plus ultra"*, that he had been the furthest south of any man onboard.

No sooner had *Resolution* arrived back in Great Britain from the second voyage than it was announced that a third was planned. Initially a Captain Charles Clerke was reported to have been chosen to command the expedition but Cook, who had recently been promoted captain, volunteered for yet another voyage. His offer was accepted and he set to, readying his ships. As tender to *Resolution* he chose the 80-foot-long Whitby-built collier, or cat, *Bloodhound*. Captain Clerke took command of this ship. William Bligh, later to become captain of HMS *Bounty* — notorious for a famous mutiny — was that ship's master. Before leaving, Cook rechristened the *Bloodhound, Discovery*, a name already famous in the history of exploration, being the same as that of Henry Hudson's ship on his fourth and last voyage on which he searched for the Northwest Passage. Perhaps *Discovery* was applied with this in mind, as one of the main objectives of this third voyage was an investigation of a Northwest Passage entering the Arctic from the west.

On his second voyage, Cook had disproved the theory that there was a habitable continent in the southern latitudes, but the cartographers still held to the theory that there was a Northwest Passage and this idea had been fuelled by Samuel Hearne's recent expedition which followed the Coppermine River through to the Arctic Ocean.

The navigable possibilities of such a passage were based on the pseudoscientific thinking that as sea water did not freeze, the Arctic ice was a product of freshwater rivers. The proponents of the Northwest Passage could not conceive of the impregnable quality of the Arctic ice.

When the Hudson's Bay Company was founded in 1670, its charter stipulated that it should attempt to resolve the possibility of a passage. But the company devoted itself instead to the very profitable fur trade. An expedition led by James Knight, a governor of the company, sailed into Hudson Bay in 1719 and was heard of no more. Hearne's expedition from Fort Churchill to the Arctic Ocean crossed no salt water or major river and this added to the reluctance of the Bay's governors to spend money on a search that, if successful, would only bring others to a territory where they had such a financially rewarding monopoly.

No indications of the entrance to a passage had been found in the east, but perhaps entry from the west held the key to success.

On 6 July 1776 Cook received his instructions, and the Royal Society, having examined all known data and read the many theories and unauthenticated reports of voyages, gave the plan for the cruise its final shape. The timetable laid out was somewhat precise in nature but Cook was consulted on it and must have approved. He was to leave Tahiti at the beginning of February 1777 and without delay sail to the coast of New Albion at latitude 45° North. This was to put him on the coast of America with Spring and Summer before him for a coasting trip to latitude 65° North "or farther, if not obstructed by Lands or Ice; taking care not to lose any time in exploring Rivers or Inlets, or upon any other account, until . . . into the before mentioned Latitude of 65°".

The reason for latitude 65° North was that it was known that it was at about this latitude that the Russian landmass bulged to the eastward and thus it must be about here that the North American continent trended eastwards if it were to join with Hearne's open Arctic waters.

Once reaching this area, Cook's instructions were to "very carefully search for and to explore such Rivers or Inlets as may appear to be of a considerable extent, and pointing towards Hudsons Bay or Baffins Bay".

The Greek pilot Apostolos Valerianos, sailing for the Spanish under the name Juan de Fuca had, in 1592, claimed to have found a broad inlet leading eastwards which he had followed for more than twenty days, between latitudes 47° and 48° North. The Royal Society gave little credence to this tale and the only reason Cook's landfall was designated at about latitude 45° North was to allow him to make repairs and take on fresh supplies after his Pacific crossing.

In due course, Cook picked up the west shoreline of America in the vicinity of Cape Foulweather on the Oregon coast. The same kind of weather for which the cape was named forced him to keep well offshore and when he eventually reached latitude 48° North, where the strait that now bears the name of Juan de Fuca is located, he passed the mouth of it in darkness. He had missed it, and the comment in his journal read: "It is in the very latitude that we were now in where the geographers have placed the pretended Strait of Juan de Fuca, but we saw nothing like it, nor is there the least probability that ever any such thing existed". His ships continued up the west coast of Vancouver Island without realizing that it was not a continuation of the mainland.

Thus on 29 March 1778 they came by chance to Nootka, an anchorage of little significance to Cook other than as a place of refuge, repair, and replenishment.

Cook came to anchor in Ship Cove, now known as Resolution Cove and remained there for a month refitting his ships, trading with the natives, till 26 April.

While the refit was underway, Cook, as was his practice wherever possible, set up his observatory ashore. He observed longitude by means of ninety sets of lunar observations all being corrected by the ship's chronometers. He also observed the variation of the compass and made tidal measurements.

Cook was a hard taskmaster as the following lines written by one of his midshipmen at Nootka attest:

Oh Nootka! Thy shores to our labours attest,
(For 30 long miles in a day are not jest)
When with Sol's earliest beams we launched forth in thy Sound,
Nor till he was setting had we compass'd it round,
Oh day of hard labour!

Captain Cook in Nootka Sound.

The reference to thirty long miles tallies with the geographical realities of a trip around the group of islands lying in the middle of Nootka Sound, the main island subsequently being named Bligh Island, with the long Clerke Peninsula forming its southwest extremity and the Spanish Pilot group lying to the westward.

This and other days of hard labour produced the sketch survey of Nootka Sound, though this name was not applied till later. Cook usually tried to apply a local name, but in this case he settled on the title King George's Sound while recording that the Indian name appeared to be Nootka. It was this name that was later applied in England.

Cook's hydrography in western Canadian waters was confined to Nootka and some coastal work and, his refit over he quickly voyaged north into Alaskan waters (crossing on 17 August 1778 the Arctic Circle, thus becoming the first explorer to cross both extremes of the globe) only to find himself barred by the Arctic ice of the Bering Sea. If there were a Northwest Passage, it was not his to find and, his ships and crews punished by the cold and fierce squalls, Cook returned south to the Hawaiian Islands which he had discovered earlier on this same voyage; there on 21 February 1779, in a dispute with the natives, Captain Cook was killed.

The main impact of his visit to Vancouver Island came after his death when the sea otter furs he had obtained in trading with the natives brought large sums of money on the China coast. The publication, in 1784, of Cook's journals and their mention of the abundance of otter, and the natives' willingness to trade

Why the British are called "Limeys"

In 1753 a British doctor published a *Treatise of the Scurvy* in which he proved that eating oranges and lemons could cure the disease. After his discovery the Royal Navy ordered all its ships to carry lemon juice on long voyages. However, in the nineteenth century, in the midst of the great flurry of Arctic exploration, lime juice was substituted for lemon. The term "Limey" was applied by North Americans to British seamen at first, later to anyone from Great Britain. When lime juice was substituted for lemon, scurvy again became epidemic; today it is known that limes are only half as effective as lemons in preventing the disease. When vitamins were discovered in the early twentieth century, and when vitamin C's effectiveness as an antiscorbutic was recognized, vitamin pills became standard issue in the Royal Navy.

furs for drygoods, led quickly to an active trading commerce on the newly found Canadian west coast.

The detailed hydrographic charting of Canada's west coast was to be left to one of Cook's young officers, George Vancouver, who as mentioned accompanied him on his second voyage and was a midshipman with Commander Clerke, HMS *Discovery*, accompanying Cook in *Resolution* on the third voyage.

Vancouver sailed from England in April 1791, with orders to sail to the northwest coast of America, to survey the coast from latitude 30° North to 60° North and to resolve two specific issues: "1st [his instructions read], The acquiring accurate information with respect to the nature and extent of any water-communication which may tend, in any considerable degree, to facilitate an intercourse for the purposes of commerce, between the north-west coast, and the country upon the opposite side of the continent, which are inhabited or occupied by his Majesty's subjects." (Though the eighteenth century language is a little obscure, Vancouver's instructions were a clear signal that his primary task was to search for a trade route through the Arctic.) The second issue was the determination of Spanish claims of sovereignty to the territory around Nootka Sound, Vancouver Island, previously surveyed by Cook.

Although Cook is credited with discovering the west coast of Canada, the Spaniards had made several prior voyages of exploration along the coast. But they had never succeeded in landing, and their reports and charts were not generally published but fell under the veil of secrecy that the Spanish authorities normally imposed on such material.

In 1775, Bodega y Quadra in command of the *Sonora*, reached the shores of Vancouver Island on 10 August, probably in the vicinity of Wickaninnish Bay on the shores of the present-day Pacific Rim National Park. He beat his way as far north as latitude 57° where the coast of Alaska trended westwards and, as passage further north presented the difficulties of running westward, he decided to turn south and on 8 September, under a favourable wind, headed towards the coast, sighting the Queen Charlottes about eight or nine leagues offshore. He did not venture closer as so many of his small crew were suffering from scurvy and he did not consider it prudent to close the land with insufficient numbers of men to manoeuvre the vessel in inshore waters.

It is not clear how much of the coast of the Queen Charlotte and Vancouver Islands Quadra saw and charted, as his records show an unbroken coast between these islands, but with no particular features. Further south very few observations were made, as by this time Quadra and his first officer Juan Pérez were also down with scurvy. They reached Monterey on 7 October and in company with the *Santiago* sailed for San Blas on 1 November, Pérez dying before reaching that port.

The Spanish planned another voyage for 1776 or the following year but in fact it did not take place until 1779 by which time Cook, with his better equipped ships, experienced hydrographic officers, and well trained crews, had made the coast. There also was a philosophical difference in the attitudes of the British and Spanish authorities in that London published charts and voyage reports, whereas Madrid followed a policy of secrecy. In a negative way this policy was so successful that it is only in recent years that some of the Spanish documents have become available for study by scholars, and the extent of the Spanish exploratory surveys has become known.

At any rate, as a direct result of Cook's voyages, the west coast trade in otter skins was flourishing by the time of Vancouver's voyage. John Meares, a British trader, had formed a company which built a post at Nootka Sound. In 1789, the Spaniards, who had claimed the whole region by reason of the, by then ancient, Treaty of Tordesillas of 1494, seized the post and several British ships that were in harbour at Nootka. The two nations almost went to war over the dispute but conflict was avoided by the signing in Madrid of the Nootka Sound Convention of 1790, by which Spain conceded British sovereignty and made an offer of restitution. Vancouver was instructed to stop off at Nootka to receive this offer in person.

First, though, he surveyed the northwest coast of America from a point one hundred and ten miles north of San Francisco up to the Straits of Juan de Fuca and Puget Sound. Then, while sounding in the Strait of Georgia, "As we were rowing on Friday morning [in June 1792] for Point Grey, purposing there

Scurvy: the dreaded disease

Between the sixteenth and nineteenth centuries scurvy was the ultimate scourge of seafaring men. Its effects on sailors were disastrous; one sufferer — when he'd recovered, which few did, — wrote: "It rotted all my gums which gave out a black and putrid blood. My thighs and lower legs were black and gangrenous, and I was forced to use my knife each day to cut into the flesh in order to release this black and foul blood. I also used my knife on my gums, which were livid and growing over my teeth. . . . When I had cut away this dead flesh and caused much black blood to flow, I rinsed my mouth and teeth with my urine, rubbing them very hard. . . . And the unfortunate thing was that I could not eat, desiring more to swallow than to chew. . . . Many of our people died . . . for the most part expiring behind some case or chest, their eyes and the soles of their feet gnawed away by rats." Scurvy's cause was the lack of vitamin C in the salt meats and other foods the ships were forced to carry for long voyages. The disease ceased to be a major problem when steam propulsion replaced sail, resulting in speedier passages; and when canned fruits and vegetables became available. On his voyages, Captain Cook managed to avoid a major outbreak of the disease by stocking his ships when possible with citrus fruits and juices.

to land and breakfast, we discovered two vessels at anchor . . . a brig and a schooner wearing the colors of Spanish vessels of war.'' The ships were on a peaceful charting mission and Vancouver joined them for a while. His journal goes on to note a discovery quickly made: ''I cannot avoid acknowledging that, on this occasion I experienced no small degree of mortification in finding the external shores of the gulf had been visited, and already examined a few miles beyond where my researches during the excursion had extended, making the land I had been in doubt about, an island.'' Vancouver continued his surveying and circumnavigated the island, naming it ''Quadra and Vancouver's Island'', known today as Vancouver Island. He sailed on to Nootka where eventually negotiations with the Spanish came to a standstill. (The question was ultimately resolved in 1795 in favour of the British.)

Having determined the insularity of Vancouver Island, and that the Straits of Juan de Fuca did not lead to a Northwest Passage, Vancouver spent the next two years completing his survey up the west coast as far as the Queen Charlotte Islands and Cook's Inlet, Alaska. By the time he returned to England, his ship the *Discovery* (not the same vessel as Cook's) had logged 65,000 miles, 10,000 more than on Captain Cook's longest voyage. His charts were the most accurate yet produced for Canadian waters and were so precise that some of them remained in use well into the twentieth century. Vancouver's accomplishments seem all the more remarkable when one considers that during the years of his Canadian surveys, the man was in extreme ill health. He died in May 1798, less than three years after his return to England.

If one approaches hydrography in Canadian waters as a kind of religious movement, then within the context of that analogy, Cook is the one true god and Vancouver his prophet. Their methods and the instruments they used would remain in service, basically unchanged, for the whole of the nineteenth century and not a little of the twentieth. Their reverence for the methodical, and their devotion to accuracy in positioning and sounding set the standard for all the generations of hydrographers to come and were institutionalized in the formation on 12 August 1795 of the Hydrographic Office of the British Admiralty. From this bureau, during the nineteenth century, scores of men — often referred to as ''Men of Admiralty'' — dedicated to the examples of Cook and

Vancouver, swarmed over the sea coasts and the major rivers and lakes of the British holdings in North America. This was an era of consolidation for the British Empire. There was little need to discover and lay claim to new lands. Now what was required were charts accurate enough to allow for colonization, the export of materials harvested by the colonials, and the import of necessary goods from the motherland.

No other single factor has played a greater part in charting the coasts and seaways of Canada than the search for the Northwest Passage. Cabot and Cartier in the east, and Cook and Vancouver in the west were expressly despatched from Europe in search of this passage, the discovery of which, it was hoped, would open a fast route from the Orient to the markets of Europe. Later, almost the whole of the Arctic was to be explored and charted solely in the quest for this route to the East.

The first of the major Arctic explorations were headed by Sir John Ross and Sir William Edward Parry (appointed Admiralty Hydrographer, 1823) who on their voyages in the first two decades of the nineteenth century discovered and charted much of the Arctic coasts and many of the islands from Baffin, west to the Beaufort Sea. But they found no through passage.

By the 1840s, through the efforts of Admiralty hydrographers and work done by surveyors in the employ of the Hudson's Bay Company, most of the northern coastline of Canada had at least been navigated if not charted. The expeditions had approached the Arctic from the east at Baffin Island, from the west through the Bering Strait, and overland from either Hudson Bay or the delta of the Mackenzie River in the Northwest Territories. No one had yet completed a nautical passage through the Arctic. So in 1845 the Admiralty, with no little sense of mounting frustration, dispatched Sir John Franklin with "explicit instructions concerning the course he should follow in order to find and to navigate the North West Passage." (Don W. Thomson)

By then it was realized that the passage was of negligible commercial value, but, nonetheless, considerable prestige would accrue from its successful navigation. Moreover, the learned societies in England welcomed and supported

Hydrographers frequently found rocks and shoals the hard way — by hitting them. This painting by Lieutenant Bedwell shows HMS Plumper *aground off Waldron Island, B.C.*

the opportunity of obtaining Arctic observations, and a considerable number of scientific instruments was stowed aboard the ships of the expedition.

Franklin himself was an experienced explorer who had been surveying in the Canadian Arctic since 1819. But on this voyage, tragedy struck, and Franklin and his two ships, the *Erebus* and the *Terror*, and the one hundred and twenty nine men of their crews vanished. The Franklin expedition was last sighted making for Lancaster Sound on 26 July 1845. It was equipped for a three-year stay in the Arctic, so that serious public and official concern was not felt until 1848, when the first search parties were organized. In his book, *Men and Meridians*, Don W. Thomson notes that "the scope and persistence of the numerous expeditions sailing into dangerous and largely unknown waters [in search of Franklin], has had no parallel in the maritime history of the world. During the decade of the search (1849-1859), 33 ships wintered in the Canadian Arctic. Hundreds of crewmen and many officers were involved in all these projects and the total cost to the British government amounted to millions of pounds sterling." And although the principal object of the voyages during this period was the search for Franklin, these expeditions added an immense store of new knowledge to the cartography of the Canadian Arctic.

By 1859 the Franklin search had for all practical purposes come to an end. A cairn was finally discovered with letters detailing the death of Franklin and the loss of his ships and men to the ice, and this period of intensive surveying of the Canadian Arctic also was concluded. Most of the "firsts" had been achieved and, in particular, the coast of the Beaufort Sea (named after Sir Francis Beaufort, Admiralty Hydrographer, 1829-55), home now to much oil and natural gas exploration, had been almost completely charted. By the way, in the course of the Franklin search, Commander R. M'Clure, R.N., and the crew of his ship, *Investigator*, finally proved the existence of a navigable Northwest Passage† and were awarded the £10,000 Admiralty prize for their discovery. By this

†Ironically, M'Clure and his crew discovered the passage while on foot; M'Clure, his second master, and seven seamen were trekking overland from their ice-bound ship seeking a route out of the bay in which their vessel had become entrapped when they found the Northwest Passage. Later, when the Franklin cairn was discovered, it became obvious that the Franklin expedition deserved credit for its prior discovery.

Investigator beset in ice.

time, however, the passage's commercial applications were worthless.

On the west coast, meanwhile, Vancouver's charts were felt to be sufficiently accurate and extensive enough to cover most contingencies. But some Admiralty survey work was conducted by Captain Henry Kellett and Sir Edward Belcher on the Columbia River, the present boundary between the states of Washington and Oregon, Vancouver Island, and the Straits of Juan de Fuca until 1848 when most of the Admiralty's ships, monies, and energies were directed to the search for Franklin.

In 1857, Captain G.H. Richards, in command of HMS *Plumper*, arrived on the west coast, initially to conduct surveys required for the settlement of a dispute involving the correct determination of a United States/Canadian border in the area now bounded by Washington state and British Columbia. Richards

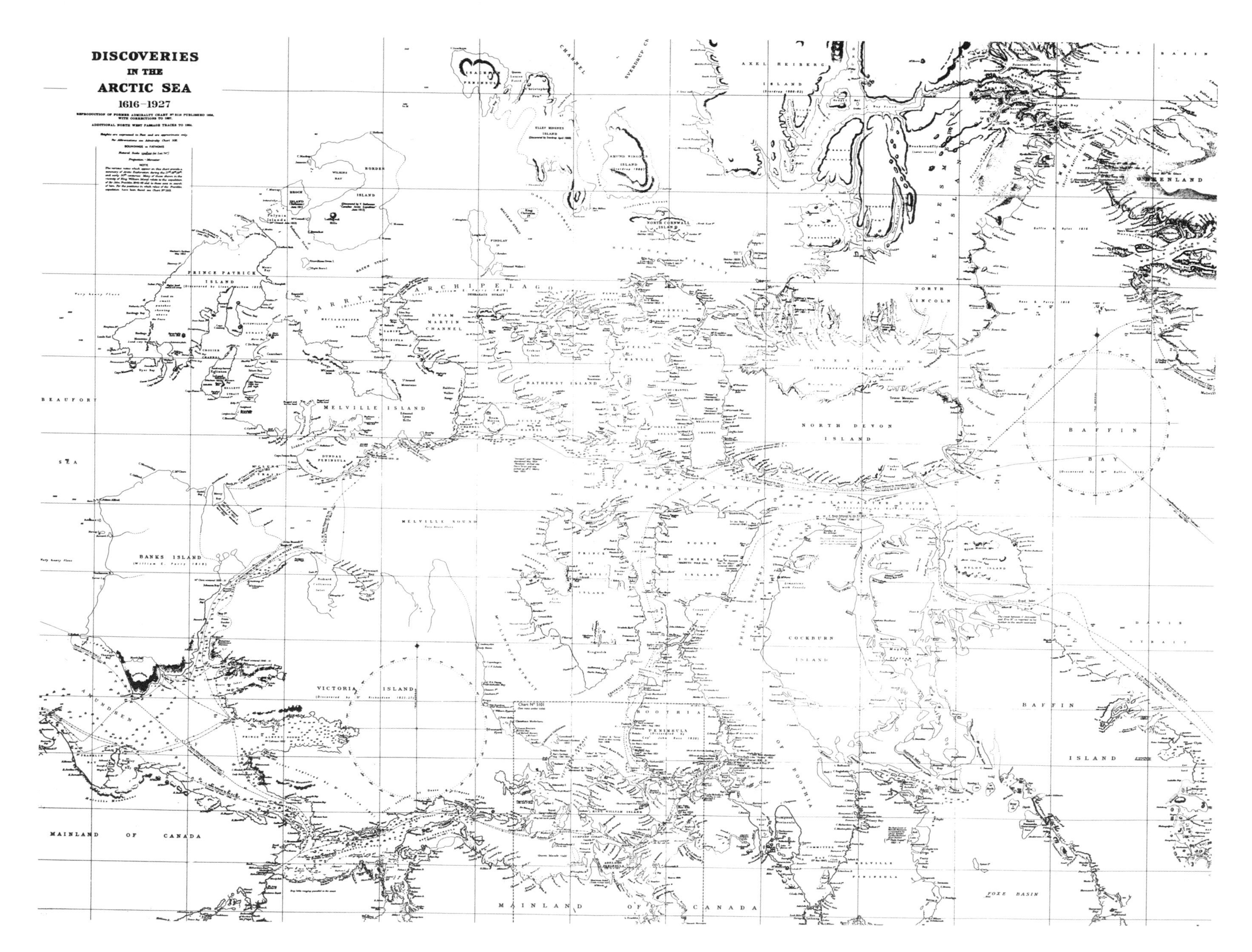
DISCOVERIES
IN THE
ARCTIC SEA
1616–1927
SOUNDINGS IN FATHOMS
NOTE
AXEL HEIBERG ISLAND
ELLEF RINGNES ISLAND
AMUND RINGNES ISLAND
NORTH CORNWALL ISLAND
GREENLAND
KANE BASIN
BROCK ISLAND
BORDEN ISLAND
PRINCE PATRICK ISLAND
PARRY ARCHIPELAGO
BYAM MARTIN CHANNEL
MELVILLE ISLAND
BATHURST ISLAND
NORTH LINCOLN
JONES SOUND
NORTH DEVON ISLAND
BAFFIN BAY
BEAUFORT SEA
DUNDAS PENINSULA
MELVILLE SOUND
BARROW STRAIT
LANCASTER SOUND
BANKS ISLAND
PRINCE OF WALES ISLAND
NORTH SOMERSET ISLAND
BYLOT ISLAND
COCKBURN ISLAND
DAVIS STRAIT
VICTORIA ISLAND
BOOTHIA PENINSULA
GULF OF BOOTHIA
BAFFIN ISLAND
AMUNDSEN
MAINLAND OF CANADA
MELVILLE PENINSULA
FOXE BASIN
MAINLAND OF CANADA

first concentrated his surveys in the contentious boundary area and then, in the 1858 and 1859 seasons, he worked around the Gulf Islands, and also in the Fraser River, Burrard Inlet, Victoria, and Esquimalt regions. He systematically surveyed around Vancouver Island in an anti-clockwise direction. Under Richards's command, a buoyage system was implemented on the Fraser River, and the first Canadian west coast "Notice to Mariners" was issued, giving details of this system.

Richards was recalled to England — he'd been promoted Hydrographer of the Royal Navy — and from 1863 to 1870 west coast charting continued under Captain Daniel Pender. Pender chartered the Hudson's Bay Company steamer *Beaver* and set to work, first in the waters south of Queen Charlotte Sound. By 1866 this area was fairly well charted and most of the remainder of his time was spent in northern waters, including those around the Queen Charlotte Islands. In 1869 Pender surveyed Burrard Inlet investigating sites for buoys, lighthouses, and beacons at the request of the British Columbia government which wanted to develop sawmill interests in the area. This was before the founding of the city of Vancouver.

By 1870 the west coast had been well opened to safe navigation, mainly through the efforts of Richards and Pender; nearly thirty years elapsed before another British survey ship was appointed for any length of time to the Pacific station.

During the first half of the nineteenth century in eastern Canada — on the Great Lakes, along the Saint Lawrence River, in the Gulf of Saint Lawrence, and on the coasts of Nova Scotia, Prince Edward Island and New Brunswick — Admiralty hydrographic surveying was dominated for forty years, from 1817 to 1856, by one remarkable man, Admiral Henry Wolsey Bayfield.

O.M. Meehan, a former Canadian hydrographer and for many years an indefatigable chronicler of Canadian Hydrographic Service history, has noted that:

> of the 215 Admiralty editions of charts issued to 1867, approximately *114* or *53%* of the total were attributed to Bayfield in whole or in part . . . After 40 years of field service he retired in 1856 and could look back with some personal satisfaction in knowing that there were few sections along the main steamer routes between Halifax, N.S., and Fort William in Lake Superior that he had not had a hand in charting.

On the whole, I think being at sea when [I am] not sick . . . is one of the most delightful places in the world.

Rear-Admiral Sir Francis Beaufort
Admiralty Hydrographer, 1829-1855

When the Canadian government founded the Georgian Bay Survey in 1883 as the first national hydrographic service, it appointed Staff Commander John George Boulton, RN, as its first chief, his initial duty was to update the Bayfield charts.

Commander J.G. Boulton, who in 1883 was named to head what eventually became the Canadian Hydrographic Service, later eulogized Bayfield in this fashion: "The Admiralty Surveying Service has produced good men, from Cook onwards, but I doubt whether the British Navy has ever possessed so gifted and zealous a Surveyor as Bayfield."

Bayfield joined the Royal Navy when he was ten years old, and first saw Canada in 1810 at the age of fifteen when his ship touched at Quebec City and Halifax. The year 1814 found him, at the end of the War of 1812, serving as an acting master with the British flotilla on Lake Champlain.

In 1816 he was recruited by Captain William Fitzwilliam Owen, senior officer commanding on the Lakes, and naval surveyor, as an assistant on the Great Lakes Survey, charting Lake Ontario and the Thousand Islands. It was here that Bayfield learned the rudiments of hydrography. Quickly. For the following year, Owen was recalled to England and at twenty-two years of age Bayfield found himself in charge of the surveys of Lakes Erie and Huron. This was almost more curse than blessing for, though the task before him was considerable, he'd been left with only one midshipman assistant and two small boats.

Nevertheless, they completed a survey of Lake Erie in 1818 and then sailed to Penetanguishene where they began to chart Lake Huron. Examples of the difficulties this work entailed are supplied by Ruth McKenzie in her monograph on Bayfield:

> Ten weeks were required to survey 45 miles of the north shore because, Bayfield explained, '. . . in that distance we have ascertained the Shape, size and situation of upwards of 6,000 Islands, flats and Rocks; the main shore too is broken into deep Bays and Coves. . . Many a night I have slept out, when the Thermometer was down to near Zero [Fahrenheit], and sometimes even below it. Yet even this was not so wearing as trying to sleep, in vain, in the warm nights of summer in the smoke of a Fire to keep off the clouds of Moschettoes which literally darkened the air.' Sometimes the surveyors and crew suffered from ague, sometimes from scurvy, and they had no medical aid.

By 1822, Bayfield was able to report that he had completed "the Survey of Lake Huron up to the Rapids of the Neepish, at the entrance to Lake George". Thus the next year he began his survey of Lake Superior. The charting of this lake, at that time almost unknown to anyone except native Indians and fur traders, required three summers' work and a further two years in England completing the charts.

In 1827, Bayfield expressed a desire to return to Canada to chart the Saint Lawrence from "the Western Shores of Newfoundland to Montreal", citing in his request his opinion that "there are few parts of the Globe in which more accidents occur to Vessels than in the dangerous Gulf and River St. Lawrence." His wish was granted and Bayfield spent the next fourteen years, headquartered at Quebec City, in the conduct of the Saint Lawrence Survey. It was a monumental undertaking and the Admiralty, recognizing the importance of the mission to the development and maintenance of their colonial interests, provided Bayfield with a ship built especially to his specifications, the 140-ton *Gulnare*.

From this vessel Bayfield and his assistants would set out in twenty-five-foot boats, six days a week and sometimes seven, from sunrise to sunset, surveying coastlines and taking soundings during the summer months. When autumn weather became too severe to permit work in the gulf, Bayfield and his men sailed upriver working in the sheltered waters around Montreal and Quebec. In the winter they compiled their plans and charts and sent them off to the Admiralty in London to be engraved. The proof sheets were shipped back to Bayfield for final inspection before printing and publication.

In 1841 Bayfield moved his headquarters from Quebec to Charlottetown, Prince Edward Island, "a harbor that had a longer navigational season than Quebec and was more central to the projected surveying activities": McKenzie. From here, Bayfield completed his survey of the east coast and the Gulf of Saint Lawrence, his final work being the survey of Halifax Harbour in 1855.

Besides his surveys and charts, Bayfield busied himself from 1828 to 1855 working on his book, *Sailing Directions for the Gulf and River St. Lawrence*, submitting each chapter, as it was written, to the Admiralty to be printed. It was published in three stages in 1837, 1847, and 1857 and finally appeared

For almost forty years Henry Wolsey Bayfield laboured as a Royal Navy surveyor in Canada and was eventually promoted admiral for his accomplishments. Despite the superiority of Bayfield's charts, the constant improvement in ships and their motive power made his soundings increasingly unreliable.

in two volumes in 1860, under the title *The St. Lawrence Pilot*. As well, a list of latitudes and longitudes compiled by Bayfield was issued in 1857 as *Maritime Positions in the Gulf and River St. Lawrence and on the South Coast of Nova Scotia*. He concluded his writings with *The Nova Scotia Pilot*, published in two volumes in 1856 and 1860.

Bayfield retired from surveying in 1856, was promoted admiral in 1867 and continued to live quietly in Charlottetown until his death, 10 February 1885 at the age of ninety. Because of the length of his service in Canadian waters and the breadth of his work, just in terms of numbers of miles of water surveyed, Bayfield's reputation, after his retirement, assumed almost mythic proportions. Because of his well known perseverance in pursuit of accuracy, and his devotion to the examples of Cook, Vancouver and the ideals of the Admiralty Hydrographic Office, the stature of a Bayfield chart approached that of scripture. Accordingly, and perhaps somewhat unfortunately in a few cases, it sometimes took a while before those in charge of developing a Canadian Hydrographic Service could be convinced of the fallibility of Bayfield charts and the need to update them to more modern standards.

In the meantime, Admiralty surveying continued on the east coast under the banner of HM Newfoundland Survey. Bayfield's command here in the Maritimes was passed first to his former assistant, Captain John Orlebar, who, until his retirement in 1864, oversaw Admiralty work in Newfoundland, the Gulf of Saint Lawrence and the Saint Lawrence River.

It was during this time that soundings for the laying of the first transatlantic cable were taken. In 1858 the terminus survey for that cable was begun in Trinity Bay, Newfoundland. Towards the end of his tenure, one of the last important surveys conducted by Orlebar was that, in 1864, for the terminus of the second cable. Captain Orlebar was succeeded in 1865 by Staff Commander J.H. Kerr, who in turn passed on the command of the Newfoundland Survey to then Navigating Lieutenant W.F. Maxwell in 1872. In 1891 Maxwell was succeeded by Staff Commander W. Tooker.

By this time, of course, the British North America Act had created, out of the Province of Canada — that is, Ontario and Quebec, Nova Scotia, and New Brunswick — the Dominion of Canada, and with the end of British rule

Fury and Hecla Strait Survey Melville Island and Little Cornwallis Island "Ships of Opportunity"

Surveying Fury and Hecla Strait in the Eastern Arctic between the northern tip of Melville Peninsula and Baffin Island, the CSS *Baffin,* like the vessels of the early Arctic explorations, sails in uncharted waters. Fury and Hecla Strait had been considered as a possible alternative route for projected oil and gas shipments from the Canadian Arctic. The survey party faced hazards of ice, near zero weather and harsh landscape as they completed a day's work.

CCGS *Louis St. Laurent*, ice breaker, and other Transport Canada ice breakers are frequently used as "ships of opportunity" for hydrographers in Arctic operations. The *Louis St. Laurent*, working off desolate Little Cornwallis Island in the central Arctic, helps a supply ship, the *Arctic Tide*, to break through to Polaris Mine on the island. On such passages, track soundings are taken and these supplement data obtained from regular surveys.

BAFFIN

2 MINUTE

in those territories, came the official, if not the actual, end of Admiralty charting in Canadian waters. After 1867 the British began gradually to withdraw their ships from charting activities in Canada except in those areas where the Empire's shipping, trade, and communications were directly involved, as in the waters of Newfoundland and Prince Edward Island (where the Newfoundland Survey maintained its headquarters in Charlottetown, even after the province's entrance into Confederation in 1873), and the coast of British Columbia.

The federal government, however, felt it was not equipped, in terms of either capable personnel or available financial resources, to launch and fund surveys of its own, and hydrography was at first accorded a low priority in the general scheme of things.

There was, though, some hydrographic work being done by Canadians at this time. The creation of the Province of Canada through the union of Upper and Lower Canada in 1841 had provided for the establishment of a Canadian Board of Works. One of the tasks assigned to this board was the building of canals. Thomson, in his writings, asserts that "the earliest recorded accounts of Canadian, as distinct from British Admiralty, hydrographic surveys are to be found in the reports and charts of provincial engineers prepared for public works purposes. The Board of Works may be regarded therefore as a significant link (during a period of general transition) between strictly Admiralty surveys and Canadian hydrographic measuring work, particularly in bettering navigation in the St. Lawrence River system." Apart from the maintenance and improvement of existing waterways and canals, much of the time of the public works' surveyors was devoted to a survey of a possible shipping route from Georgian Bay to Montreal by way of the French and Ottawa Rivers. But this fantastical project never came to fruition.

It wasn't until 1882 that the Canadian government realized the serious need for a qualified hydrographer capable of conducting a complete, modern resurvey of at least its Great Lakes' shipping lanes. Unable to secure the services of such a one in Canada, the Dominion turned, not surprisingly, to the British Admiralty. A request for aid in establishing a Canadian hydrographic survey was forwarded by the Minister of Marine and Fisheries to the Admiralty.

The immediate impetus for this request was the wreck during a storm, on

14 September 1882 of the passenger steamer *Asia*, with a loss of some one hundred and fifty lives. This was not the first major shipping accident on the Great Lakes. But it was the largest to that time, in terms of fatalities. It was final evidence, together with a long list of previous marine casualties, particularly in Georgian Bay — the hub then of Canada's shipping industry — that the Bayfield charts of the 1820s could no longer serve the needs of the new steamships.

Steam power, in the form of wooden hulled paddlewheelers, had come to the Great Lakes during the War of 1812. By 1838 some ships were driven by propellers. It was in the 1870s that the character of Great Lakes' shipping changed radically with the introduction of deep draft, iron hulled steamships — the freighters and the carriers. These vessels were much faster and capable of closer inshore navigation than the large sailing vessels of Bayfield's day. Georgian Bay with the major ports and shipbuilding centres of Owen Sound and Collingwood was treacherous; its jagged shoreline and many uncharted shoals began to take their toll. The rising number of shipwrecks, culminating in the *Asia* disaster, did not go unnoticed by the public, the press, and finally the politicans who felt the pressure of newspaper headlines, editorials, and the displeasure of both their constituents and the business interests of mining, shipping, and the railways — all of them dependent in one way or another on the provision of accurate charts of Georgian Bay.

Following the wreck of the SS *Asia*, investigations and enquiries were held. It was then decided to approach the British in hopes that a "man of Admiralty" could be seconded to resurvey Georgian Bay to modern standards.

In February, 1883, the Hydrographer of the Royal Navy instructed Staff Commander William Maxwell of HM Newfoundland Survey to travel from his winter headquarters in Charlottetown to confer with the Minister of Marine and Fisheries in Ottawa. Ascertaining the present survey requirements of the federal government to be in the shipping centres of Georgian Bay and Lake Huron, Maxwell immediately recommended to the Admiralty that a man be sent from London. On 13 August 1883 Staff Commander John George Boulton arrived in Ottawa. The Canadian Hydrographic Service, under the name of the Georgian Bay Survey, funded by the government of the Dominion of Canada, was born.

Boulton was forty-one in 1883, an experienced naval officer and hydrographer who had served the Admiralty since he was fifteen. He had sailed as assistant to Maxwell on the Newfoundland Survey from 1872 until 1881, at which time he had returned to England with his family so that he might take his examinations for the pilotage of first class ships.

Following successful completion of those, he had been posted to a survey on the west coast of England. Dissatisfied, he had applied several times for a return to Canadian waters. He had been denied, but, perhaps having heard rumours of what was afoot in Canada, Boulton worded his last request to include mention of his interest in surveying Georgian Bay, should the Canadians decide to initiate such work. Boulton's desire, considered together with his previous service under William Maxwell, the man recommending the secondment of an Admiralty surveyor to Georgian Bay, resulted, in July 1883, in Boulton's appointment to head the Georgian Bay Survey.

Upon his arrival in Ottawa, Boulton soon discovered that he was to be very much on his own. The only guidelines given him were that he should confine his surveys to the main steamship routes between Owen Sound in Georgian Bay and Sault Sainte Marie in the Northern Channel of Lake Huron, and that he should adopt Admiral Bayfield's previously charted shorelines in his surveys. This latter was intended by the government as a cost-saving move, but Boulton was quick to recognize that Bayfield's work would prove inadequate for the needs of the new lake carriers and he convinced the ministry to allow the time and money necessary for the resurvey of the coasts.

At that point it became a matter of determining where to begin the survey. To this end Boulton travelled to Collingwood and spent a week talking with sailing masters, pilots, and shipping authorities about uncharted shoals, rocks, reefs, and about the amount of lake traffic in the bay. Taking all this information into consideration, plus the fact that the United States Geodetic Survey had accurately positioned the Cove Island lighthouse in the eastern entrance to the bay, Boulton left Collingwood for the port of Killarney, in the eastern entrance to what is called the North Channel. Here, with a few hired workers and a rented fishing tug, he laid a surveying base line and extended a "triangulation network" southwest towards Cove Island lighthouse.

It was the foundering of the steamship Asia *(berthed at Midland, Ontario) in 1882 with the loss of more than one hundred lives that led directly to the formation of the Canadian Hydrographic Service. Christy Ann Morrison, pictured here, was one of the two survivors of that tragedy.*

Bayfield *was the first survey ship to enter full time service with the Canadian hydrographic fleet. Purchased in 1884 for Boulton's work on Georgian Bay, the former tug remained in service on the Great Lakes until she was replaced by the second* Bayfield *in 1902.*

By the end of August, a promised government ship had not materialized, so Boulton took it upon himself to charter a tug, the *Ann Long*, for forty-five days, the rest of the season. She was an uncomfortable little 72-footer, and not equipped at all for survey work, but she did the job and goes down in history as the first ship used by the Dominion government for hydrographic work.

Boulton, despite the handicaps of having no hydrographic ship and no trained assistants, was both pleased to be on his own in Canadian waters and eager to prove his worth to the Canadians. Consequently, by October, settled in an office in the West Block on Parliament Hill, he had gathered enough information to begin writing the first chapter of *The Georgian Bay and North Channel Pilot*. He spent the winter preparing field sheets for future surveys, and drawing up specifications for a new, government survey steamer.

Included in Boulton's mandate was the responsibility both to recruit and train Canadian hydrographers. To meet this requirement, and because he needed help desperately, Boulton recruited William James Stewart in March 1884. Stewart was a first graduate and gold medallist from the Royal Military College at Kingston, and a future chief Canadian hydrographer himself. In the spring of that year, the government purchased, for $15,000, its first hydrographic steamer, the *Edsall*, a former tug. For another $4,000 the *Edsall* was remodelled and refitted for survey work and renamed *Bayfield* in honour of the man who had pioneered hydrographic work on the Great Lakes.

For the next few seasons, Boulton and Stewart concentrated on extending their survey north and west along, in Boulton's words, "the present trade route, not feeling justified in putting the country to the expense of surveying water which at present, a vessel has no inducement to pass." Boulton, however, was not unaware of the possible needs of future hydrographers in the area, and noted in that same report that "should minerals be discovered, or any industries spring up, it will be an easy matter to extend the survey over the particular locality, and with this contingency in view, the centres of the main triangulation stations have been marked by broad arrows into the rocks or iron bars driven into the soil."

By 1886, the first chart of the Georgian Bay Survey had been published by the Admiralty, a second assistant to Boulton had been hired, and the *Bayfield* had undergone another $5,000 worth of additional remodelling and refitting. She could now carry two smaller survey boats and a total ship's complement of twenty-three personnel — including the three hydrographers. Also that year Commander Boulton established the first permanent Canadian bench mark in the form of a monumental block of stone near Little Current on Manitoulin Island. Of this he wrote, "the levelled top being 6 feet 9 inches above mean summer surface level of the water, these figures have been engraved upon the top of the stone to serve as a permanent bench mark for future reference and comparison."

Boulton was a man in a hurry, but he was also the quintessential hydrographer in his willingness to concede that sometimes the only way to "get it right" was to take one's time. In 1889 he wrote, about a section of the northeast coast of Georgian Bay between Byng Inlet, where the SS *Asia* foundered in 1882, and the Limestone Islands, that:

> work on this portion of Georgian Bay must necessarily be slow, for a more broken-up coast line it is impossible to conceive, and the same up-and-down character of the bottom is extended to the sea for two or three miles in the shape of many dangers very hard to find by the ordinary methods of hydrographical surveying . . . the only way to navigate a coast of this exceptional character is to adhere exactly to the leading marks given on the charts and sailing directions, and not to make too free with this uneven bottom, though the chart may show more than sufficient water. Sounding in the dark waters of the northeast coast of Georgian Bay where a rock with only 6 feet on it cannot, at times, be seen, is only groping about in the dark at the best, and although our lines are sometime only 100 yards apart — not a great distance, when the enormous expanse of the lakes yet unexamined is considered it sometimes happens that no indication of a rock is given with the lead. I mention this fact to show that *hydrographical work cannot be hurried excepting at the risk of leaving out dangers, entailing the loss of the reputation of the officer in charge, and perhaps of valuable lives.*

In 1890 the North Channel survey was completed. Boulton reported that that season was "the finest yet experienced" and that "a vessel can now pro-

Quadra's launch waiting for the tide at Hole-In-The-Wall, British Columbia.

ceed from Owen Sound to Sault Ste. Marie, a distance of 200 miles, over recently surveyed waters.'' After completion of the North Channel survey there was still enough good weather left in the season to allow the *Bayfield* to steam to Parry Sound where Boulton and the rest of his party surveyed channels leading to the harbour. Again Boulton proved to have an eye for the future. He wrote of this stretch of water that ''The general outside traffic along the coast, the numerous islands and occasional inside channels are inducing tourists to make it a summer resort.'' Indeed, that part of Georgian Bay would provide, in not too many years, inspiration for painters of the Group of Seven and has grown steadily over one hundred years to represent the epitome of ''cottage country'' for many residents of the cities of Ontario; it is also one of the province's most popular cruising grounds for which the first recreational chart was published.

In June 1890, the Canadian Pacific Railway (CPR) steamer *Parthia* ''touched'' a shoal in the waters of Burrard Inlet in British Columbia. The incident was reported to the pilotage authority and the area investigated by HMS *Amphion*, stationed at the Royal Navy base at Esquimalt. At the time, Britain was still trying to maintain responsibility for marine surveying in the waters off all Commonwealth countries, but active hydrographic ships were not always available to examine all newly discovered shoals in all parts of a vast and far-flung Empire. In the case of Burrard Inlet, the federal government sent a request for survey to the British government which replied in the affirmative, but with the suggestion that perhaps the matter could be most expeditiously handled by someone from Commander Boulton's Canadian operation.

Early launches with mast and sail.

Apparently, with the North Channel survey completed, Boulton felt sufficiently secure in his Great Lakes' work to allow Stewart to leave the Georgian Bay Survey, in the summer of 1891, to travel to the west coast for this resurvey of Burrard Inlet and Vancouver Harbour. That was the first saltwater survey in the history of the Canadian service. The next saltwater survey on the west coast would not occur until 1906 — fifteen years later. And Canadian hydrographers would not commence sounding on the east coast until 1905.

As the last two maritime provinces — British Columbia on the west coast,

Prince Edward Island on the east — joined Confederation in the early 1870s, the Dominion government was pressured to assume responsibility for the charting of its own shores. But the simple fact of the matter was that it could barely afford the men, ships, and money for hydrographic operations on the Great Lakes, let alone on two sizeable seacoasts. Accordingly, requests to the British Admiralty for surveys were made on an "as needed" basis, with the understanding that the Canadian government would pay half of the costs of each venture.

On the west coast, during the decade after British Columbia's entrance into Confederation in 1871, Admiralty hydrographers made reconnaissance surveys, at the request of the Canadian government, of inlets suitable for a terminus for the CPR. But it wasn't until 1898 that an Admiralty hydrographic ship was again stationed on a regular basis at Esquimalt. This was the HMS *Egeria* and she remained on the coast until 1910. Her reasons for being there were several. In 1899 *Egeria* spent four months engaged in ocean sounding along the route of the proposed Pacific cable from Canada to Australia. The following year, she began resurveying the east coast of Vancouver Island, bringing charts up to modern standards in a response to increased shipping in the area, a result in part of the activity associated with the Klondike gold rush of those years.

In 1906 the focal point of surveying on the west coast switched from the populated southern areas to the relatively virgin, northern coast of British Columbia. The Grand Trunk Railway was pushing its way west and adequate charts had to be produced to open a new terminal to shipping from the Orient. Prince Rupert was the eventual location of that terminal, and the Canadian service, lacking an expert saltwater hydrographer, had to make do with a party headed by a former Dominion Land Surveyor — G.B. Dodge, who admittedly had served with the Admiralty's Newfoundland Survey. Port Simpson, however, was first intended as the site of the new harbour, and it was from there that *Egeria*'s last Canadian surveys began. By 1909 they were completed and in 1910 HMS *Egeria* was decommissioned and sold. The Admiralty hydrographic presence, on Canada's west coast, went with her.

On the east coast, during the same 1867 to 1908 period, the Admiralty maintained the Newfoundland Survey at Charlottetown, and from 1908-12 in

G. B. Dodge, a land surveyor who had acquired some hydrographic experience with the Royal Navy, was assigned the task of surveying Prince Rupert Harbour, the chosen west coast terminal of the Grand Trunk Railway. He was assisted by Cdr. P. C. Musgrave and H. D. Parizeau who completed the survey which became the first all-Canadian chart of the west coast.

Even after the establishment of the CHS, the British Admiralty retained responsibility for some west coast surveys. HMS Egeria *was based at Esquimalt for twelve years, decommissioning on station in 1910. Captain Parry and his officers are shown in a sylvan setting in the Naval Dockyard. Note the ship's dog in the left foreground.*

Halifax. Commander Maxwell, to whom Boulton had been assistant, maintained command of the survey until 1891. His successor, Staff Commander William Tooker, was involved in one of the last British surveys conducted for Canadians, the 1892 survey of a stretch of Anticosti Island, the scene of many shipwrecks over the years. Captain J.W.F. Coombe succeeded Tooker in 1908. The last survey of the Newfoundland Survey was of Saint John's Harbour and approaches, completed 31 October 1912; the chartered steam yacht, *Ellinor*, used on that survey, then departed for the West Indies, the Halifax office of the Newfoundland Survey was closed, and an illustrious part of Canada's hydrographic history came to an end.

The Admiralty, however, did conduct two further significant surveys in Newfoundland waters. In 1932 and 1933 HMS *Challenger* carried out work, initially intending to "survey a route inside the islands from Indian Harbour in the south to Cape Chidley in the north". It was an overly ambitious goal, still not accomplished. In two years — including a winter survey — Commander A.G.N. Wyatt completed the survey of Nain. The price to the Admiralty is indicated by Challenger Rock on which the ship was badly damaged.

The last Admiralty survey was of Saint Lewis Inlet, also on the Labrador coast, by HMS *Franklin* under Lieutenant Commander C.W. Sabine in 1939.

Meanwhile, in 1892, the Georgian Bay Survey under Commander Boulton was drawing to a close. The first volume of Canadian sailing directions, *The Georgian Bay and North Channel Pilot*, was published. An act of the federal parliament placed both tidal observations and hydrographic surveys under the direction of the Chief Engineer's branch of Marine and Fisheries. And Commander Boulton set out on his last field season. By October, Boulton could state without doubt and not a little pride that:

Navigating Lieutenant W. F. Maxwell was one of several Royal Navy officers named to the command of the Newfoundland Survey after Boulton had been seconded to duty on the Great Lakes.

> Two more seasons should complete the survey of Georgian Bay and the North Channel of Lake Huron. The total number of nautical miles of coast line surveyed has been about 2,560: the boat sounding amounts to 8,224, while 9,203 miles have been sounded in the ship. The cost of this has been approximately $188,000, giving an average value of $73 for each mile of coast surveyed. The United States have about the same quantity of lake coast line as Canada; their survey was commenced in 1841 and finished in 1881, their total cost being two and three quarter million dollars.

On 12 April 1893, Boulton officially handed over his command of the Georgian Bay Survey to Stewart, his assistant since 1884. Upon his return to the Admiralty Hydrographic Office in London, he was appointed naval assistant to the Hydrographer and served in that position until his retirement in 1898 with the rank of captain.

That same year, Boulton returned to Canada to live in Quebec City where he died in 1929 at the age of eighty-seven.

William J. Stewart was the first Dominion Hydrographer though the title at that time was Chief Hydrographic Surveyor. He had been one of the first graduates of the Royal Military College at Kingston and was the silver medallist from his graduating class.

In 1893, Boulton's successor, William J. Stewart, following the orders of Chief Engineer William Anderson, set out in the *Bayfield* with two new assistants — one of whom, Frederick Anderson, was to be Stewart's successor — to complete the Georgian Bay Survey. The Chief Engineer himself went, with a rented boat, to resurvey the Bay of Quinte on Lake Ontario. This expedition was required because of the increase in shipping on the bay due to the completion in 1889 of the Murray Canal. The charts for the bay resulting from the resurvey were not published, however, until 1900, as a result of delays encountered in the process which required Canadian charts still to be printed and published by the Admiralty in England, a situation that was not to change until 1903.

By 1894, the Georgian Bay Survey was — except for the middle of the bay which would not be sounded until 1964 — completed. It had taken eleven years and more than $215,000 to produce thirteen charts and a volume of sailing directions. But the demands of the complicated shorelines and the many islands and channels of Georgian Bay had been excessive, particularly for a survey that had no more than three hydrographers at work at one time. Accordingly, Stewart felt confident in writing that in the future "with the possible exception of Lake Superior, none of the other lakes will take anything like that amount of time or money".

This remark was especially appropriate in light of the statement made only four months later by William Anderson, the Chief Engineer, and Stewart's immediate superior, in his annual report of January 1895: "the hydrographic survey of the Georgian Bay and North Channel which was most urgently required, having been completed, it has been decided to continue work on the remaining Canadian waters of the Great Lakes. The use of deeper draught vessels and the increasing speed of steamers make the demand for reliable charts very urgent." Thus was born the Great Lakes Survey which was to occupy Stewart and his assistants for the next eight years.

The survey seasons of 1895-7 saw the *Bayfield* busy charting the shores of Lake Saint Clair, Lake Erie, and Lake Huron. But in 1898 she returned to Parry Sound for a resurvey of the main steamer channel there, with a plan to improve that passage's aids to navigation — mainly range lights and spar buoys. This action was a result of the establishment that year of a line of large freighters set to carry goods to and from the new terminus in Parry Sound of the Ottawa, Arnprior, and Parry Sound Railway. Later that season the *Bayfield* returned to a survey of the rest of Lake Huron, and she remained there for the next three seasons, the last of which, 1901, was under the command of Stewart's first assistant, Frederick Anderson.

Stewart himself spent the summer of 1901 on the *Frank Burton*, a chartered steam tug, surveying the waters of the southern portion of Lake Winnipeg. Since the erection, in 1898, by Marine and Fisheries, of two lighthouses on the lake, increased shipping traffic on these uncharted waters had demanded that some hydrographic surveys be undertaken.

Work continued on Lake Winnipeg for three years, the last two under the supervision of Stewart's assistants as he had returned to the Great Lakes. In the end, the Winnipeg survey was perhaps most important because the chart produced from it was also the first *Canadian* chart from *Canadian* surveys, the "first official occasion when results from a Canadian survey were not forwarded to the Admiralty for engraving and publishing": Meehan. It is now known as *Chart 6240, Red River to Berens River*. Unfortunately, the Chief Engineer later had to report that "the demand for this chart has been exceedingly small."

In 1902 Stewart, the lone hydrographer on the *Bayfield* since both his assistants were on Lake Winnipeg, commenced his long-anticipated survey of Lake Superior. Having pronounced the aging *Bayfield* "totally unfit" for this survey and "being obliged to take all fixes alone", Stewart understandably was limited in the amount of work that he was able to accomplish that season. But the following year, though still without his assistants and forced to resort to hiring "some transient students", Stewart was equipped with a new ship, the larger, more powerful *Lord Stanley*. Immediately rechristened *Bayfield*, this vessel allowed him to begin sounding off the more exposed and windy northern shores of Lake Superior as far west as Thunder Bay.

That season of 1903 was the last of what was called the Great Lakes Survey. It was also the last field season for William J. Stewart. In 1904, by authority of Order-in-Council (P.C. 461), the government of Canada amalgamated the hydrographic operations of the Department of Publics Works, the Department

Frederick Anderson succeeded Stewart as chief hydrographer and it was during Anderson's tenure that the service was officially renamed as the Canadian Hydrographic Service. Anderson held his master's papers for small craft and during his years as head of the service was always referred to as "Captain."

of Railways and Canals and the Ministry of Marine and Fisheries into a Canadian hydrographic service.†

The hydrographers who came from Public Works could claim a longer history than those from the Great Lakes Survey. They could trace their origins to 1841 when an Act of Union created the Province of Canada out of Upper and Lower Canada (Ontario and Quebec). In August 1841 a Board of Works was established for the new province and its responsibilities included marine surveys and dredging of the Saint Lawrence River and the Great Lakes.

Under the British North America Act of 1867 the Board of Works became the federal Department of Public Works with responsibility for improvements to Canadian ports and rivers. In 1896 a hydrographic survey unit was established and its major task was surveying the Saint Lawrence below Montreal to plan for the dredging required to provide a 27½-foot channel.

In May 1879 the Department of Railways and Canals was calved off from Public Works and in 1884 established its own hydrographic unit. From then until its amalgamation with CHS in 1904 it carried out surveys on the Richelieu River, Lake Saint Louis and the lower Ottawa River.

A second Order-in-Council of 1904 went on to state:

> all hydrographic work in the Dominion should be under the management and control of a Chief Hydrographic Surveyor having his headquarters permanently at Ottawa. . . It will therefore be seen that Mr. William J. Stewart has a very large and long experience as a Hydrographic Surveyor upon the Great Lakes, and has also some salt-water experience where currents were very strong and tides of great range. . . The Minister therefore recommends that Mr. Stewart be appointed Chief Hydrographic Surveyor of Canada.

†From the time of Boulton's Georgian Bay Survey in 1883 there had been, in fact, a Canadian hydrographic service. The 1904 Order-in-Council broadened the service's responsibilities, but not its basic function, and changed its name to Hydrographic Survey of Canada. Unofficially the new name was generally corrupted to "Canadian Hydrographic Survey" and it was not until 1928 that the present name, Canadian Hydrographic Service, was officially adopted. Despite the name changes the purpose and function of the organization has remained constant; throughout this book we use the term "Canadian Hydrographic Service (CHS)" to designate the continuing traditions of Canadian hydrography even though, in some instances, the use of this term will prove anachronistic.

So Stewart, who preferred the title "Chief Hydrographer", assumed control of the reconstituted hydrographic service. Coincidentally, in June of that same year, the Admiralty, beleaguered by requests for surveys and resurveys from British dominions and colonies around the world, requested of the more self-governing ones such as Australia and Canada that they institute their own marine survey departments and "conduct hydrographic surveys along their own coasts". The Admiralty also expressed its expectation that Canada would be the first to do this. And of course it was right. In fact, we'd anticipated the request by a good four or five months. Really, Canadians had been attempting to tend to their own hydrographic needs since the hiring of Boulton in 1883. True, he had been a *British* officer, a "man of Admiralty", but Canadians had paid both his salary and the expenses of his Georgian Bay Survey. The fact that Boulton was British seems now only a logical extension of the legacy handed down from Captain Cook, the father of hydrography as it is known today, the man who built upon and consolidated the work of those French, Spanish, and Portuguese mariners who for hundreds of years had navigated and charted the waters of Canada.

In the end, it was Boulton and his Canadian student, William J. Stewart, who set the pattern for the shape of the Canadian Hydrographic Service whose centenary we now celebrate. Meehan in his unpublished history of the service's early days, commenting on the importance of the first ten years, those of Boulton's Georgian Bay Survey, lists Boulton's achievements in terms of the charts and the *Sailing Directions* published under his signature. However, Meehan then goes on to state that:

> What was more significant were the *techniques* and *practices* of hydrographic surveying Commander Boulton passed on to his successors, practices that were 'norms' for the Hydrographic Service until the gasolene launch was introduced in 1904, and the . . . echo-sounder in . . . 1929.

It is this legacy of "techniques and practices" and their further development in the twentieth century that we will next examine.

The Measures

Measures not men

Lord Chesterfield (1694-1773)

Measures, not men, have always been my mark

Oliver Goldsmith (1728-1795)

Not men, but measures

Edmund Burke (1729-1797)

EASURES NOT MEN" WAS A PREDOMINANT MOTIF OF THE eighteenth century Age of Reason. The philosophy behind the motto was that the truly scientific, rational and therefore successful man would put his faith in the objective observation of facts, of "measures", rather than the subjective or merely intuitive perceptions of "men". It was this credo that led, among other things, to the invention of the instruments and methods used by Captain Cook — instruments and methods that transformed the *arts* of navigation and marine cartography into the *science* of hydrography.

In Canada, the means and measures of Cook's day remained in use, relatively unchanged, almost to the middle of this century. But the attitude that produced the instruments and methods of the eighteenth century, with its insistence on "measures, not men", was eventually tempered. The first influences for change were the romantic impulses of the nineteenth century and of the Canadian frontier; later came the humanism of the twentieth century; and finally, the radical technological developments of the past thirty years, developments which in their ability to raise more questions than answers, have shaken any lingering belief in the infallibility of science.

It's thus important at the beginning of this chapter on instruments and methods that one keep in mind the fact that the hundred-year history of the Canadian Hydrographic Service is one not only of measures, of changing technologies and methodologies, but also of concurrent changes in both the attitudes of those people responsible for the implementation and use of new develop-

The greatest mountain range on Earth is not the Himalayas or the Rockies, the Andes or the Alps. The widest and deepest chasm is not the Grand Canyon of the Colorado. The broadest plain is not the steppes of Russia or the Great Plains of North America. It may come as a surprise to landlocked minds, but the greatest mountains, chasms, and plains all lie beneath the oceans, a grandeur unseen and until recent times unmapped and unmappable.

John Noble Wilford
The Mapmakers, 1981

Our knowledge of the sea is infinitely small. We work at sea knowing next to nothing about that medium. We use the sea to carry our ships, to carry sound-waves, to propagate energy of all kinds, without having a clue how sound-waves behave and travel in sea-water, without being able to predict the effect of wind on the sea or the behaviour of our ship in a swell. In short we claim to know the depth of the sea and the presence or absence of wrecks in a certain area . . . without knowing anything at all about the medium we use to collect this information. This is, in fact, a queer situation born in a time when the oceans and their boundaries were considered to be non-changing and featureless; when mankind had a high opinion of its scientific knowledge . . .

Commander W. Langeraar
Royal Netherlands Navy

ments in surveying and charting, and the attitudes and needs of the Canadian people.

At the end of the chapter it will be evident that no matter how sophisticated the hardware that is brought to the task, it is no substitute for the dedication of men devoted to ideals of truth and perfection, to the idea of "getting it right". In that spirit, and in fairness to Edmund Burke, we should perhaps place the remark quoted above within its proper context:

> Of this stamp is the cant of, Not men, but measures; a sort of charm by which many people get loose from every honourable engagement.

In later chapters we'll take a closer look at the men of the Canadian Hydrographic Service, men whose work, upon which the lives and safety of so many people depend, can with little argument be characterized as an "honourable engagement".

For now, the history of the CHS can be detailed first in terms of its instruments and methods in the field, then the development of chartmaking in Canada, and finally, the history of the ships used by Canadian hydrographers in their work.

The science of hydrography is concerned essentially with measurement. Unlike the work of many other scientists, however, the hydrographer's task involves the measurement of the largest object of which man has firsthand knowledge — the Earth itself, or more precisely, the waters that cover two-thirds of the Earth's surface.

Consider for a moment a section of the Rockies and the Alberta foothills. Imagine a passenger aboard a sophisticated flying carpet that floats some few thousand feet above ground at, say, Calgary. It is his task while aboard the carpet to make the observations necessary to produce a contour map of the region over which he is to fly. Oh yes, the carpet passenger has one major problem to contend with — clouds quite impenetrable to the naked eye obscure the ground beneath him. Now, how — precisely — does he go about producing his map of the Rockies from Calgary west to the coast?

Most basic tool in the navigator's kit is the magnetic compass. In its earliest form it was simply a magnetized needle inserted into a cork or plug of wood and floated in a bowl of water; when the needle stopped oscillating it pointed north. Later the needle was suspended in air, or fluid, upon a pivot and still later a card, printed with the cardinal points, was attached to the needle and swung with it so the mariner could read the ship's heading directly from the card.

It is obvious that before he takes off he will have done some fairly detailed planning. And he'll probably have had to invent some new instruments or improve and amend existing equipment; a yardstick may prove helpful when he steers the carpet close to the peaks that project through the cloud cover, but when the ground lies thousands of feet or metres below . . . what then?

Basically, hydrographers sailing about the oceans of the world and scurrying about the Earth's surface have been in an analagous situation. They've had to do some complicated thinking about the job of charting the waters of the Earth, and they've had to invent sophisticated instruments for dealing with the job. The first of these, and perhaps the simplest, was the compass.

Before the hydrographer can begin to take his measurements and before the mariner can begin to read the hydrographer's chart, they both need to know "which way is up". They need to orient themselves, find out in which direction they're going or would like to go; and this involves first of all discovering in which direction North lies†. We saw in a previous chapter that the intro-

†The convention of placing North at the top of maps was adopted in relatively modern times. Medieval maps placed the East — that is, the Orient — at the top, a device that gave rise to the word "orient".

duction of the magnetic compass allowed the first European navigators to make their famous voyages of discovery. This same magnetic compass, with which former Boy Scouts and Girl Guides are familiar, appears a relatively uncomplicated instrument. A suspended iron needle is attracted by the magnetic pull of the North and South Poles. Unfortunately, for the purposes of long distance navigation and accurate chartmaking, the matter is not so simple.

Some authorities credit Columbus with discovering on his crossings of the Atlantic that his compass did not always point exactly North; the attribution is debatable and probably false. But whether it was Columbus or another mariner who first noticed the difference, it was early in the age of European exploration that the distinction between "true" and "magnetic" North was made.

True North is synonomous with the North Pole — the point at which all meridians of longitude converge; magnetic North is the place — a shifting position, to be sure† — at which the needle of a magnetic compass points. Magnetic North is about 1,600 kilometres south of the north pole, midway between Bathurst and Ellef Ringnes Islands.

It wasn't until 1928 that Canadian hydrographers began to use a different kind of compass altogether. This was the gyroscopic compass, first invented in Germany in 1906. This compass relies on the gravitational pull of the spinning Earth to align a similarly spinning set of wheels on board a ship in such a way that an attached compass card *always* indicates the position of true North. For navigational purposes, this was a considerable breakthrough, particularly for those working in Arctic waters where ordinary compasses, in close proximity to the magnetic pole, are virtually useless.

The first gyroscopic compass was installed on the *Acadia* during the 1928 survey season when the ship was working along the north shore of the Gulf of Saint Lawrence. The compass was a great success; the experience led the

†The Poles' attraction for the compass needle is caused by the rotation of the planet's molten interior. The mass of this body of liquid rock shifts as the Earth tilts about its axis and circles the sun; its influence on the compass needle is correspondingly altered. In addition to the variations in magnetic North, a ship's magnetic compass is affected by a number of other factors. Metal built into a ship's hull or fittings can disturb a compass's accuracy, and a magnetic compass is always more accurate when a ship is sailing generally in a northerly or southerly direction than when heading east or west.

then chief hydrographer, Captain Anderson, to exult that it was "a most satisfying and valuable acquisition to the vessel's charting navigational equipment." More importantly, Anderson recognized immediately that this instrument would prove invaluable for "magnetic variation investigation and general survey work in Hudson Bay should this ship [*Acadia*] be detailed for work in northern waters" — which it was, the very next year.

The only derogatory remark on record about the gyro compass was made by J.U. Beauchemin, the hydrographer-in-charge of the *Acadia*. He reported that "I have given all my spare time this summer in studying the different parts and mechanism of the Master Compass, but I must admit that I would be at bay to make any repair in case of emergency." Though he spoke in jest, his comment was indicative of the kinds of changes that these instruments of the future would demand in the type of personnel required for their operation and maintenance.

O.M. Meehan, writing in the 1960s, remarked that "the installation of the first gyroscope compass on the survey ship *Acadia* marked the commencement of a transition from the older order of surveying to one with [electrically and eventually] electronically operated units." This was the first step in the long march of a coming revolution in hydrographic surveying.

Though these new compasses were relatively expensive (almost $6,000 in 1928), their value to the work of an organization dedicated above all else to accuracy was such that they were soon installed on the other major ships of the service, the *Lillooet* (in 1930), the *Wm. J. Stewart* (1932) and the *Cartier* (1934), and are now standard equipment on all hydrographic vessels and major ships.

The gyroscopic compass or "gyro" uses a spinning wheel to keep its compass card constantly aligned in a north-south axis. A child's spinning top demonstrates the same principle.

Now our hydrographer has a device that can tell him which way is up — or rather, North. Now what's required is a means of deducing exactly *where* he is on the surface of the Earth.

For the hydrographer, this matter of positioning is of the utmost importance. If it is his job to provide information for mariners on underwater hazards to navigation — shoals, rocks, wrecks, and other perils to ships — it follows

Don't do more than is wanted; don't do less than is needed.

Traditional maxim of hydrographers

that he must be able to convey the whereabouts of these dangers and to indicate safe routes for navigation. The means by which these positions are expressed on a chart are the coordinates of latitude and longitude.

For those of us who need not cope with latitude and longitude daily, the concept may seem confusing. It need not be.

In their study of astronomy some five thousand years ago, the Babylonians determined that the Earth circled the sun once every 360 days.† From this conclusion, came the convention of dividing a circle — the ancient astronomers perceived the Earth to be a sphere, a three-dimensional circle — into 360 equal parts or degrees. Thus was born the concept of longitude; if the Earth is viewed from either the North or South Pole with the pole centred, the 360 lines that may be drawn from the centre, representing the 360 degrees of the circle, become also the meridians of longitude.

Later, the Greeks noticed that during the span of a year the sun "migrated" from north to south as it circled the Earth.†† Each year, on about 22 June, the sun reached its northernmost progression, and on about 22 December the southernmost. Similarly, the Greeks noted, on about 22 March and 23 September of each year the sun seemed to cross a point midway between the most northerly and southerly extremities. Drawing a line about the Earth perpendicular to the poles and passing through this midpoint, they labelled it the Equator, and drew parallel lines through the sun's northernmost and southernmost limits, calling them the Tropic of Cancer and the Tropic of Capricorn respectively.†††

Thus the Greeks were able to divide their spherical Earth in half by means of the celestial equator which in terms of latitude was designated as zero degrees. The distance from the equator to either the top or bottom — the poles — of

†Though their calculation was not precise, the Babylonians' determination of a year's duration did not affect the use to which they put it.

††The Greeks were in error, of course: the sun did *not* circle the Earth but just the reverse. This misconception, like that of the Babylonians, did not alter the effectiveness of their theories.

†††"Tropic" comes from the Greek word *tropos*, "to turn", for, according to John Noble Wilford, "it seemed that at that time the sun stopped and reversed itself, or turned about."

Using a theodolite, hydrographer R. J. Fraser, who was later to become Dominion Hydrographer, makes observations of the Sun's angular elevation to determine latitude on an Atlantic coast survey in the 1920s.

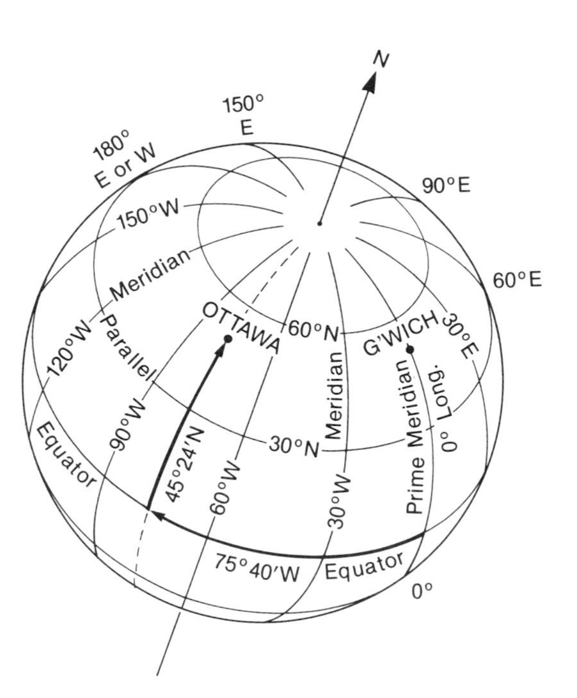

the sphere was thus one quarter of 360 degrees, or 90 degrees. By scaling off the globe, North and South from the equator, they divided the Earth into hemispheres, each graduated from 0° to 90° — parallels of latitude.

The means for describing one's position on the surface of the Earth were in place; in both vertical and horizontal planes, imaginary lines circled the globe — meridians of longitude from North to South, parallels of latitude from East to West — forming a grid. Using the intersections of the grid lines as reference points a navigator could say with a modicum of accuracy where he was at any moment. But it was, at best, a theoretical position and many years were to pass before the instruments were available to permit him to determine his position with real accuracy.

By the time of Boulton and Stewart, at the beginning of the twentieth cen-

Dead reckoning is no more than a jumble of pieces, all of them wrong to varying degrees. Hence the results obtained in this way can bear no resemblance to reality unless by some miracle the errors cancel one another out . . .

Jean Randier
Marine Navigation Instruments
quoting from M. Leveque's
Guide de navigation

Taking a round of angles with a theodolite (above) a hydrographer calls his readings to an assistant who takes notes.

tury in Canada, the methods and instruments for establishing latitude and longitude with precision were considerably more than one hundred years old.

The theodolite, an instrument used in land surveying for determination of vertical and horizontal angles, was invented sometime in the 1500s; the sextant was invented in 1731; the ship's chronometer, necessary for an exact determination of longitude, came into use about thirty years after that. It wasn't until World War II that the three devices began to be supplanted by electronic positioning equipment.

The sextant was a refinement of instruments in use by navigators since at least the late Middle Ages. All of these earlier devices, the astrolabe, the cross-staff, and their descendants, the quadrant, octant, and back-staff had one purpose — to measure the angle between the sun at noon or the north star at night and the horizon. This angle, when compared with tables of the sun's deviation from its apparent path around the Earth's equator, gives the navigator/hydrographer a relatively accurate determination of his latitude.†

The second device required for the computation of position by a Canadian hydrographer at the turn of the century was one which we now take for granted — the chronometer, or simply an accurate timepiece. This was needed for the calculation of longitude.

Prior to the invention of the nautical chronometer navigators estimated their position, east or west, by what was called "dead reckoning".†† Using this method, the navigator set sail from one port knowing, from his chart, the distance to the next landfall to the east or west. Relying on his ability to calculate and maintain, through the use of his compass and quadrant or sextant, his lateral movement across the surface of the Earth, the navigator then verified his position by checking the speed of his vessel, using the ship's log.

†The deviation of the sun from its apparent path, known as the sun's declination, is detailed for the navigator in a handbook titled the *Nautical Almanac*, a copy of which is carried by all ships.

††Some authorities believe that the phrase, "dead reckoning", derived from the abbreviation, "ded. reckoning", or "deduced from reckoning", inscribed on his charts by the ship's navigator when indicating his daily position at noon.

Figuring the average speed of a ship under sail was problematic at the best of times and obviously this method of navigation was flawed. With the increase in maritime traffic in the sixteenth, seventeenth and eighteenth centuries the irony of "dead reckoning" was proved as the practice resulted in more and more shipwrecks with a consequent loss of many lives and valuable cargoes. In fact, the problem reached such proportions that in 1714 an act of the British parliament offered a prize of £20,000 to anyone who could invent a means of accurately determining longitude.

The dilemma facing those eighteenth century scientists is expressed well by author John Noble Wilford in *The Mapmakers*:

> The longitude of a place is its angular distance east or west of 0° longitude, the prime meridian. (On most maps today this is an imaginary north-south line drawn through the Royal Observatory at Greenwich, England . . .) The longitude of New York City, for instance, is 73° 59' 31'' W., that is its angular distance west of Greenwich. If you could slice a long wedge out of the Earth along the meridians at Greenwich and at New York, and if the wedge reached to the axis of the Earth, the exposed angle at the axis would measure 73° 59' 31''.
>
> But longitude can be more readily understood and reckoned as a function of time. In fact, meridian means midday. Since the Earth rotates 360° every 24 hours, it turns 15° every hour, 15' every minute. When it is noon at Greenwich it is an hour before noon at 15° W, six hours before noon at 90° W, and midnight on the opposite side of the earth at 180°. . .
>
> Since time and longitude go hand in hand it should be a simple matter to determine the position of a place east or west of the prime meridian. Simple, that is, if you know simultaneously the local time and the time at the prime meridian.
>
> . . . Nowadays, a mapmaker can get an accurate time fix by short-wave radio or by having a precision clock set to the prime meridian time — Greenwich Mean Time. Neither was available in the seventeenth century, and this was the crux of the problem.

Finally, in 1759, an Englishman, John Harrison, invented the first successful marine timekeeper — it was actually Harrison's *fourth* model, his first having been produced in 1735 — eventually winning for himself the £20,000 prize.

Where permanent and prominent landscape features do not exist, hydrographers are compelled to build shore beacons as marks on which to sight their sextants. Once the beacon's position is established, a metallic "rock post" is driven into the ground as a permanent reference for future surveyors.

A multiplicity of prime meridians

From the time the ancient seafaring peoples of the eastern Mediterranean began to push out from shore in boats, mariners have based their charts on a prime, or first meridian (numbered as 0°) from which all others are numbered sequentially eastward or westward. During the burst of exploration in the fifteenth and sixteenth centuries each maritime nation chose its own prime meridian — the Dutch chose the meridian that ran through Amsterdam, the British London, the French Paris. But this system created unnecessary complications for the navigator. Late in the nineteenth century the meridian through Greenwich, England, was internationally adopted as prime meridian. And, since longitude is a function of time, the time at the prime meridian was adopted as Greenwich Mean Time. This name has been changed in the last few years to Co-ordinated Universal Time.

James Cook, using a version of Harrison's clock on his second and third voyages on HMS *Resolution*, proved its practical value and accuracy beyond a doubt by successfully matching the results of calculations based on the chronometer with those resulting from another method, invented around the same time as Harrison's timepiece. That involved painstaking and time-consuming sightings of the moon. Because accurate chronometers were very expensive and remained in short supply throughout the nineteenth century, the calculation of longitude by means of lunar observations remained in common use up until about 1900. But by the time of Stewart, ships' chronometers were the rule rather than the exception, and part of any survey party's regimen was the manual winding of the clocks once a day, every day, at the same time. This was done with a sense of ceremony approaching that accorded a religious rite.

The compass, the sextant, and the chronometer are essentially the tools of a navigator; they are not those of the surveyor. But the hydrographer, to be effective, must combine the expertise of both the navigator and the surveyor. This is to say that while the hydrographer's prime concern may be to measure the minimum depth of water over a shoal, if his work is to be of use to other sailors, he must also be able to show on his chart exactly *where* the shoal is in relation to safe navigation channels.

Since the time of Captain Cook hydrographers have known that to produce an accurate chart they must begin ashore. In the past the shore-based work began with the establishment of a base line; selecting a promontory and building a beacon, the hydrographer would then use his theodolite to sight on the sun or stars to determine the beacon's exact location. Then, again using the theodolite, he measured the angle from his first beacon to a second established some distance away; by this means he could establish true North. Finally, he measured — on relatively flat ground, with steel tape; on rocky terrain, with trigonometric calculations — the distance between his two beacons. This precisely plotted base line permitted him to extend his triangulation network along the shoreline.

With a series of shore beacons visible from the ship, and with each of the beacons exactly positioned by latitude and longitude, the hydrographer could return aboard to begin his soundings. Using his sextant, held horizontally, the hydrographer could accurately position each sounding in relation to three or more of the shore stations.

This was the method in the past. Since 1905 when the Geodetic Survey of Canada (GSC) began its field operations, CHS has had to establish fewer and fewer base lines or observe astronomical positions; to date, GSC (with the cooperation of CHS) has measured and marked some 126,000 geodetic control points across the country, so many that at no place along Canada's extensive coastline is a survey ship today more than forty miles distant from one or more points. Indeed, so reliable and extensive is this geodetic data base that since 1961 CHS has had to measure virtually no base lines.†

Though relieved of the tedious task of establishing an accurate base line for each survey the Canadian hydrographer still proceeds, although to a lesser degree, by triangulation. A good explanation of triangulation is that given by Mary Blewitt in *Surveys of the Sea* (1957). She writes:

> Accurate surveys are based on a rigid land triangulation, that is to say that the coast and coast line are precisely mapped so that the sea bed can be charted and the details positioned accurately in relation to the land. The fundamental idea of triangulation is that, knowing the angles of a triangle and the length of one side, the lengths of the other two sides can be calculated. Two points, *A* and *B*, are selected and the distance between them measured. This is the base line. A third point, *C*, visible from both *A* and *B* is then chosen and the angles of the triangle *ABC* measured with a theodolite. When the length of *AC* and *BC* have been calculated the three points can be plotted [on paper] so that they are correctly related to each other. This original triangle is extended by others until the area to be surveyed is covered by a framework to which subsidiary data, such as the position of conspicuous features, are added.

While the Canadian hydrographer these days is relieved of the painstaking labour of establishing base lines, he continues to measure distance. Now, however, he calls upon electronic aids. The tellurometer, the instrument now most used to measure distance accurately, was invented in South Africa in 1956. The

†This statement must be qualified: hydrographers occasionally still measure base lines, mainly in the Arctic, but now they use satellite navigation receivers rather than physical taping to do so.

At more than one hundred thousand places throughout the country, the Geodetic Survey of Canada and the CHS have placed permanent markers, accurately positioned as to latitude and longitude. When a surveyor begins a new survey, or re-surveys an area, he starts from these known locations. This marker is one of hundreds positioned by the International Boundary Commission during the establishment of the boundary between Canada and the United States during the years 1908-25.

How the "beavers" helped

In the early days of the CHS, before permanent rock posts were established on shore, starting a triangulation network often involved a good deal more than setting up a theodolite and taking sights. On *La Canadienne*'s 1912 survey season in Lake Superior, for instance, the hydrographers had to cut a line through tall timber to take their sights. They were assisted by crew members, a motley group picked up from various waterfronts and Quebec forestry camps. The French Canadians, all experienced woodsmen, watched in fascination as the English seamen hacked away with more gusto than expertise. "*Castor, castor* (Beaver, beaver)," they muttered, and scrambled to safety as the trees fell in unpredictable directions.

CHS began to use it the following year. As John Noble Wilford explains:

> The tellurometer can operate in daylight or darkness; it is a two-way microwave system in which a modulated radio signal travels from a master transmitter to a remote unit, where it is retransmitted back. The distance between the two points is derived from the measured transit time of the radio signal and the known velocity of radio waves.

Tellurometers now used by the CHS can accurately measure distances in excess of thirty kilometres. They are heavy, about twenty-five kilograms with battery, and cumbersome, but their weight is offset by two great advantages: first is the instrument's ability to measure distance over rugged terrain, say from one hilltop to another (which is why its weight is considered a drawback — somebody has to carry it up the hill); second, the tellurometer permits the hydrographer to extend his survey control from a much longer base line, which tends to increase accuracy. Furthermore, the base line is not only longer, it is much simpler to establish; using the older method the hydrographer had to compensate for many inherent inaccuracies: the unevenness of the ground, the tension in his steel rule or chain, temperature changes which affected the length of the rule. Today, a tellurometer or geodimeter, used in combination with other advanced instrumentation, can be fed the temperature, humidity, barometric pressure and other data, and *display* the distance, corrected for all the variables.

In 1930 the first aerial photographs were commissioned by the CHS. That year photos were taken from the air, over the east coast, of the area between the Aguanish River and the Strait of Belle Isle, and on the west coast, over Barkley Sound, Vancouver Island. Thomson, in *Men and Meridians* notes that from that year, "air photography has since continued to be an indispensable aid to the charting of Canadian waters and, over the years, various refinements in this special service have been developed, including the provision of photographs of specified areas, taken both at low and high tide." Now, when a hydrographer begins to plan his season's survey, among the things he gathers first are the aerial photographs of the coast or shore to be charted.

He also puts his hands on whatever surveys of the area have been done previously. That most of Canada's southern waters have been charted at least once

The tellurometer is a surveying instrument for measuring distances electronically. The "Master" unit transmits a radio signal which is received by a "Slave" unit and retransmitted back. By a system of phase comparison, the distance between the units can be very accurately measured. In the Arctic, (top right) aluminum beacons are easily transported by the workhorse of the Arctic, the helicopter.

To establish shore survey control, hydrographers build beacons which are ultimately used to position the ship or launch by sextant angles. These beacons vary greatly in height depending on the distances from which they have to be seen. Beach driftwood was a favourite building material.

in the last hundred years is an obvious advantage that today's hydrographer has over his predecessors. These earlier surveys are especially useful when the hydrographer is beginning a resurvey of some coastline. Resurveys bring charts up to the higher standards required by the ever increasing draft of modern shipping, and the more precise degree of positional accuracy obtainable with electronic positioning equipment. Also, in some areas the contours of sand bottoms shift about under influence of currents and tides; periodic monitoring of minimum depths is required in such areas. In other cases new industries are established which are dependent on transportation by water of raw materials and manufactured products. While a small-scale survey may have been sufficient for passage through a channel, a larger-scale chart is required when a port to service a mine or a pulp mill must be entered. In short, a ship manoeuvering to anchor, or to tie up at a berth, requires more navigational information —

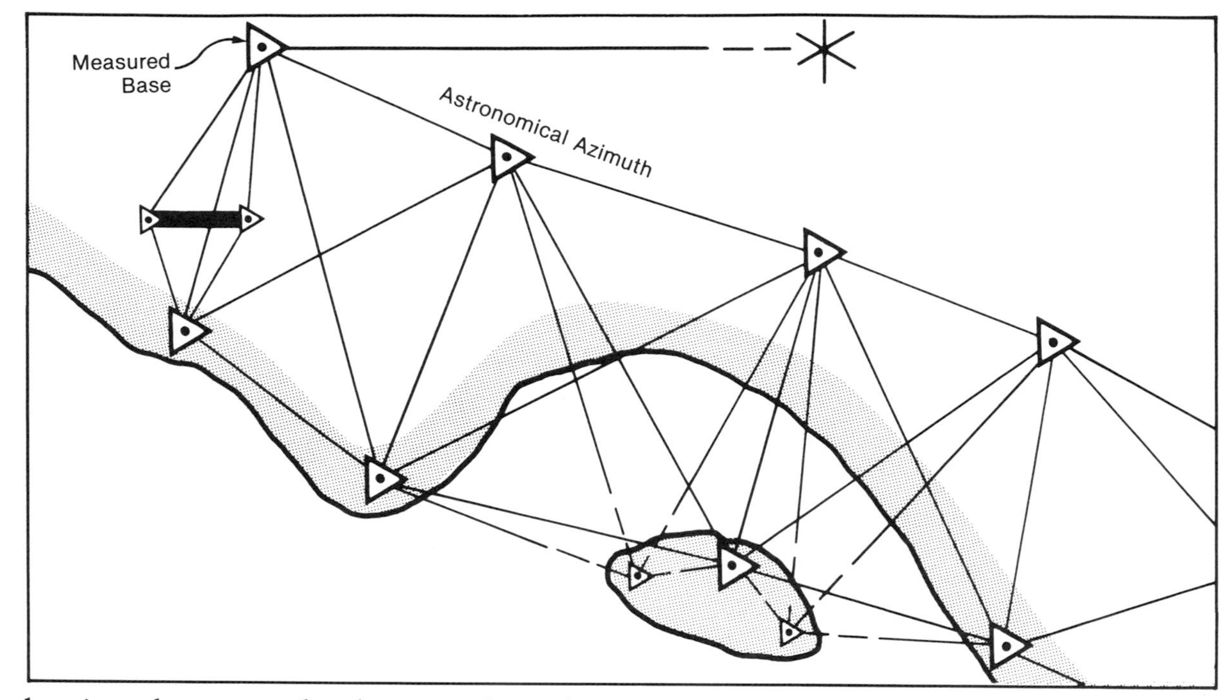

Before a hydrographer puts to sea to sound the depths he must accurately map the adjoining coastline so that his observations afloat may be accurately charted with respect to the shore. A common method of land mapping is by a technique known as triangulation, shown here diagramatically.

that is, a larger-scale chart — than the same ship merely negotiating a passage through a channel.

In the early 1930s, the United States Coast Guard was experimenting with radio acoustic ranging (RAR) for positioning offshore soundings. Meehan notes that "although not as accurate as visual fixing, it was a great improvement in positioning soundings beyond the limits of the horizon." So much so that it attracted the attention of Henri Parizeau who eventually sent the wireless operator, J.A. Nesbitt, from the west coast survey vessel *Wm. J. Stewart*, to the United States to investigate the system. RAR involved the use of three stations on shore. The position of each was very precisely observed in terms of latitude and longitude, and to each was attached a hydrophone, or underwater microphone. The crew of the survey ship at sea, out of sight of these three stations, would explode a bomb in the water at the position they wished to plot. Marking the time of the explosion and the time that the sound was received by each of the three station hydrophones, which time of reception was relayed by radio back to the ship, and knowing the speed at which sound travels through water, the position of the explosion could readily be calculated.

Upon Nesbitt's return, the men of the *Wm. J. Stewart* had time for only one test of RAR that season. In it, they successfully positioned a buoy some fifteen miles offshore. The next year, 1940, they conducted eight more trials. But because of the war and the consequent departure of radioman Nesbitt, further tests of RAR were discontinued.

Ironically, however, it was this war that would eventually do more for the development of electronic positioning devices than the twenty years of on-again, off-again experiments which it had interrupted.

Since the early 1950s the CHS has been experimenting with, developing, and using a wide variety of electronic positioning systems. All of them were based initially on the principles of RADAR (radio detection and ranging). A system developed specifically for defence purposes during the war, RADAR uses electromagnetic waves to detect and locate objects out of visual range.

One of the first of the RADAR-based systems, designed for navigational positioning, was Decca. It was initially used on the survey vessel *Kapuskasing* in 1955 but has been now largely replaced on all CHS ships by newer systems such as ARGO, Hydrodist, Hi-Fix and MRS which are specifically survey systems, portable and can be sited ashore to give the best geometry throughout the survey area. Loran-C is another generally used navigational system on the Pacific and Atlantic coasts, and the Great Lakes. The physics and logistics behind each of these varies but the basic principle remains the same.

Transmitters are erected on shore positions determined either from previous geodetic surveys or by the usual methods of triangulation. Once in place, these transmitters can operate as a single beacon (range bearing — of limited use) or in pairs (range/range — allowing for a position fix) or in threes (hyperbolic — allowing for the use of more than one sounding vessel). They may be either manned or unmanned and permanently or temporarily placed to allow the survey ships, by means of onboard receivers (referred to in the early days, with some suspicion, as the "black boxes"), to fix *accurately* and *quickly* the positions of their soundings without having to make visual sextant observations of the beacons erected on Canadian shores by hydrographers since the time of Cook. In Boulton's day they made their beacons "from driftwood on the beach". Later survey parties carried their own lumber, sometimes to the top of impossibly steep cliffs. Now beacons are still erected, sometimes with

Henri Parizeau, for many years Pacific regional hydrographer, was instrumental in the 1930s in introducing underwater acoustic positioning techniques to Canadian hydrography. However, World War II ended the experiments and war related research eventually produced the superior RADAR and other electronic methods for fixing position at sea.

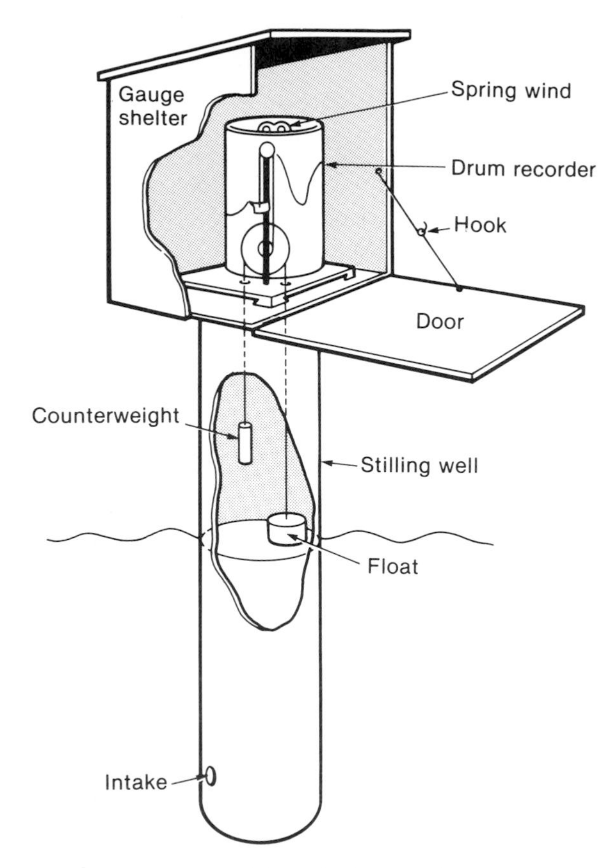

As the tide rises or falls, the float and counterweight move the wheel which drives the pen on the rotating drum recorder leaving a graphic record of the tidal height through the period of the record. As the times of high and low waters vary each day, several days' record can be obtained on one graph.

the aid of helicopters if the location is particularly daunting, but most often an electronic transmitter replaces the traditional flag.

The deployment of electronic positioning devices was useful not only for those hydrographers working out of sight of land but it also was a welcome boon to those surveying Arctic waters where often the lack of high ground or any easily recognizable terrain made the sighting of shore beacons especially difficult.

Of major importance to these hydrographers now working in the Arctic and on the open seas of the Atlantic and Pacific coasts — in the Hibernia oil fields off Newfoundland, for example — are the even more sophisticated, newer, long-range systems of ship positioning made possible through a marriage of satellite and onboard minicomputer technology. Known by a variety of names, the most common to the CHS is BIONAV, developed by the Bedford Institute of Oceanography (BIO) in Nova Scotia. These systems are valued for their extended range and their high degree of accuracy.

All the electronic positioning systems have contributed to the hydrographer's quest for accuracy in charting but, perhaps more importantly, they have eliminated much of the time-consuming drudgery encountered by hydrographers of the past in the measuring of base lines, the establishment of triangulation networks, and the sextant positioning of ships and launches taking soundings. In recent years, all of the time saved is especially significant to those surveyors working the open waters of a brief Arctic summer or fighting the vicious cold of a polar shelf expedition.

In the end, this electronic hardware allows the hydrographer to spend less time establishing control and therefore more time sounding. The electronics also mean he can sound in mist or fog, and usually with only two persons per launch rather than four.

Before any hydrographer begins sounding he must establish the lowest level to which the surface of a given body of water can usually be expected to drop. In freshwater lakes and rivers, the low-water mark will change slightly from year to year either from natural causes such as increases in rainfall, or drought, or for man-made reasons such as dams or canals. In salt water, of course, and in freshwater tidal areas such as the Saint Lawrence River as far upstream as Montreal, the water level changes constantly with the tide.

Information on tides, water levels and prevailing currents is obviously necessary for hydrographers attempting to measure and record depths. But more importantly, and certainly in a more general way, these data are required for the purposes of all navigators and mariners, both commercial carriers and pleasure boaters. The need for a systematic accounting of tidal, water level, and current information was recognized very early in the history of the service. The work began, modestly, in 1890 with the establishment of two observation posts complete with tide gauges on the Atlantic coast of Nova Scotia; the season's work cost less than $2,000. William J. Stewart, in 1891 on the west coast, installed the first Pacific tide gauge on the CPR wharf in Burrard Inlet.

Tide gauges in those days were not much more than staffs of wood with numbers painted on them and, because tides fluctuate considerably every day, the gauges required a man on a full-time basis to watch them and to record the levels of the water. In Burrard Inlet, in 1891, the tidal information gathered during the season was passed on to the CPR engineers at the end of the summer. The lowest level noted was used by Stewart as a datum on which to base all his depth sounding readings for that season.

On the Great Lakes and the Saint Lawrence River, the Department of Public Works had since the 1840s been keeping an intermittent record of water levels and currents in the major harbours and shipping lanes. But Commander Boulton in his annual report of 1891 suggested that, in the future, this function be expanded in scope and that information on the levels of waters in the lakes be made available to hydrographers and navigators alike in the form of markings placed on stones along the shore. He wrote that "as long as we have to rely only upon the fickle memory of the oldest inhabitant there will always be an element of uncertainty as to whether the waters of the lakes are subject to temporary fluctuations, or are steadily lowering their level. I would therefore respectfully suggest that datum stones be erected, say at Collingwood, Sarnia, Port Colborne and Kingston [and], that agents [of the Department of Public Works] at the ports mentioned, be instructed to note the height of the water at least once a day during the season of navigation."

In 1893, the Canadian Tidal and Current Survey was first established under Doctor W. Bell Dawson. This office was responsible for tidal and current sur-

Everyone pitches in

No matter how large a survey crew there are never enough hands to accomplish all the work assigned; each member turns to and does double duty. On a 1933 survey on the south coast of Baffin Island a CHS crew was taking tidal measurements in addition to soundings. Cribs were built, towed offshore and sunk with rocks into which measuring staffs were planted. Tides in the area averaged some thirty-three feet, equivalent to a rise or fall of more than one and one half inches per minute; the staffs required frequent — almost constant — monitoring and a hydrographer could not be spared. By careful placement of the cribs, the staffs were placed in such a way that they could be observed from shore. Several times each hour the party's cook, Jack Chester, would place down his soup ladle or meat cleaver, step out the door of the cookhouse and observe the water height through a small telescope. He would jot down the figures and then return to his stove.

W. Bell Dawson, founder of Tidal Surveying in Canada.

veys on the Canadian seacoasts (Pacific, Atlantic, Hudson Bay, and the Gulf of Saint Lawrence). The work presented two main problems for Doctor Bell Dawson: one was the development of hardware for measuring tides and currents; the other was the dissemination of the information gathered.

Don W. Thomson notes in *Men and Meridians* that "for some years . . . the federal government failed to provide funds for the publication of Canadian tables of tidal predictions. Dawson, accordingly, was forced to arrange as best he could, and from port to port, for the printing and circulation of this hard won information." Dawson persisted, however, and by 1901 the Canadian government was printing and distributing its own tide tables.

In 1924, the Tidal and Current Division was formally transferred to CHS. It was under the aegis of the service that this division later made its most significant breakthrough in terms of the publishing of tables and the expansion of services to include even parts of the Arctic Ocean, an expansion that coincided with the 1957 International Geophysical Year.

Canadian Tide and Current Tables, prior to 1967, were compiled and published in Canada, but the majority of the predictions were supplied by the Liverpool Observatory and Tidal Institute in England. These tables are predicted from thousands of hourly readings taken over the course of the year along with astronomical information. The complete set of predictions required for the Tables was compiled in Canada for the first time in 1967. Since the advent of digital computers, the process is much streamlined, reducing what once required months of labour to a computing task calculated in terms of hours.

Digitization has also had an effect on the devices used to *measure* tides and currents. Though there was progress over the years in the design of tidal gauges, leading from staffs in the water to relatively sophisticated graphic recording mechanisms, the limited storage capacity for the information gained by the machines required that graphs be changed and the recorder manually reset on an almost daily basis. With the introduction of digital technology and the increasing storage capacity of first, computer punch cards, then tape and discs, and now microchips, tidal gauges could be left unattended for longer and longer periods of time. Now as much as two years can go by before it's necessary to retrieve a gauge record though the gauges themselves still require regular checking as insurance against mechanical failure or accidental damage.

Freshwater level gauges of a mechanical, self-registering, or automatic, type began to appear on the Great Lakes about 1900. "These were the famous Haskell gauges and were installed by engineers of the U.S. Lake Survey. In 1906 three of these types of gauges were in operation by the [Canadian] Public Works Department": Meehan. In 1912, the "automatic gauges" division was transferred to CHS and in 1928, the division was renamed, Precise Water Levels. Technological development of freshwater gauges has been similar to that of tide gauges and current meters, though not so spectacular, as the demands placed on these recorders are not so great, the fluctuations in the water levels being so gradual over the course of a season.

. . . When the fourteenth night was come, as we were driven up and down in Adria, about midnight the shipmen deemed that they drew near some country; and sounded, and found it twenty fathoms: and when they had gone a little further, they sounded again, and found it fifteen fathoms. Then fearing lest they should have fallen upon rocks, they cast four anchors out of the stern, and wished for the day.

Saint Paul
Acts of the Apostles, 27: 27-9

Now, his preliminary work done, his position fixed, his datum established, finally, the hydrographer in the field comes to the central task of his mission:

> It is difficult to say that any one step in the construction of a chart is more important than another, as each is necesssary for the completion of the whole, and an error anywhere may cause a disaster; but if any particular item *is* to be picked out, perhaps the sounding should rank in the highest place.

These words were written in 1882 by Admiral William James Lloyd Wharton of the Royal Navy in his book, *Hydrographical Surveying*, a forerunner of the modern *Admiralty Manual of Hydrographic Surveying*. Wharton's manual appeared in four editions and in the last (1920) the description of sounding was exactly as it was in the first edition of 1882. Indeed, it could just as well have been written in 1782:

> The ordinary main plan of sounding is thus [Wharton wrote]. The boat proceeds in straight lines in a direction, of a length, and at a distance previously decided on, with a man in the bow constantly sounding . . . The pace at which the boat may go, and the necessity, or not, for stopping at the casts, will depend on the depth of water and the capacity of the leadsman.

Arctic tidal gauge, Resolute Bay 1957.

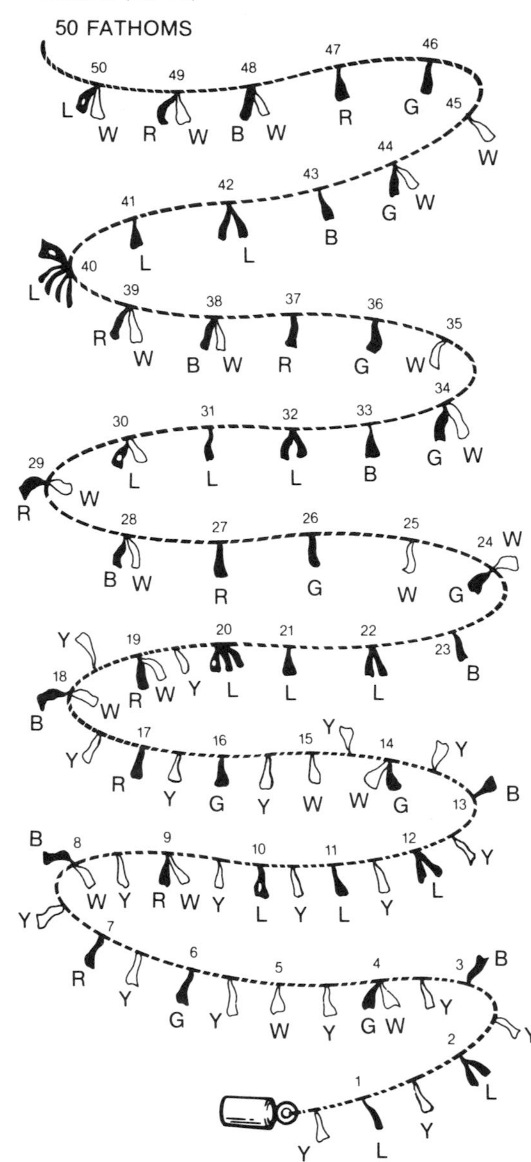

Since man first put to sea in boats he has sounded the water — that is, measured its depths — by the simple means of lowering a measured line and noting the depth at which its weighted end comes to rest on the bottom.

In other words, the method of *sounding* the depths, by means of lowering a lead weight to the seabed, remained relatively unchanged from before Cook until the introduction of echo sounders in the 1930s.

Similarly, the equipment used by the hydrographer to *record* soundings remained basically unchanged from the days of Commander Boulton who listed the field hydrographer's inventory in an 1890 lecture: "The officer takes away in his boat a small sheet [the "boat board"] of the points on the portion of the shore he is to sound . . . He also takes a sextant, station pointer [invented in 1774 and used to transfer horizontal position control angles, taken by sextant readings from the boat, on to the hydrographer's field sheet], protractor, tracing paper and pencils, not forgetting his pipe and baccy, if a smoker."

Nowadays, the field sheet, as Boulton called it or boat board as it has become known in the CHS, is plastic which is less liable than paper to distortion or water damage; the boat's position on the pre-determined sounding course is fixed by electronic positioning rather than by sextant; and on most ships almost all of the information gathered is translated into digital form and fed into an onboard computer. But all these recent innovations appear almost minor achievements when compared to the invention of the first echo sounders.

Prior to that time, hydrographers, and mariners in general, were restricted to the use of a leadline for discovering the depth of the water. In its simplest form, the lead was a fourteen-pound weight attached to a line with graduated markings on it. On some casts the underside of the lead was coated with clean, sticky tallow so that the leadsman, besides being able to determine the depth of the water from the markings on the leadline, could, upon retrieval of the lead, ascertain the nature of that bottom: that is rock, sand, mud or shingle.

The kind of bottom he sails over is important to a mariner for several reasons. If a master wants to bring his ship to anchor the nature of the bottom is of extreme importance; in strong winds a ship's anchor can drag on smooth rock or poor holding ground, and set her aground or cause her to collide with another ship in a confined anchorage; on an uneven rock bottom the hook may foul, and the number of fouled anchors that have had to be cut loose and abandoned on ocean floors is countless. On the other hand an anchor tends to bite deeply into mud or sand, holding the ship firmly within the radius of its cable, and can be hoisted without undue trouble.

Similarly, the nature of the ocean bottom is information of value to a fisherman. Certain species of fish favour specific habitats; knowing the kind of bottom beneath his hull often helps direct the fisherman to the kind of catch he wants. In addition, a knowledge of the bottom can often help prevent loss of equipment, such as nets, that foul or tear on rocks.

It's apparent then that the early leadsman's interest in the ocean bottom and the depth of water over it was no idle concern; it was based solidly on considerations of safety and commerce. An experienced leadsman could sound depths up to six fathoms without stopping the ship or boat; but depths beyond that, particularly those encountered in deep-sea soundings such as were required for the laying of the first transatlantic cable in the 1850s, necessitated holding the ship in one position as long as six hours for one sounding.

Eventually, by the time of the Georgian Bay Survey, adaptations of an invention, created originally to assist in deep-sea soundings were being used on ships sounding waters over six fathoms deep. In 1890, Commander Boulton described the use of such a machine on Georgian Bay in this fashion:

> Where the depth does not exceed about 24 fathoms the ship [the *Bayfield*, the first steamer commissioned for Canadian hydrographic work] steams steadily on at about 5½ nautical miles per hour. The sounding machine, with a lead of 25 to 40 lbs. weight attached to it, is hauled out by a traveller, on a wire rope, to the bow of the vessel. It is detached from the traveller by a trippingline when the cast is wanted. The line travels through the hand of a man aft and at a depth of over twenty fathoms the lead would be fifty or sixty feet astern of the vessel before striking the bottom. An experienced and attentive sounder easily notices the slacking up on the line, which is then brought to the steam winch and hove up . . . The interval between the soundings is regulated by an ordinary timepiece with a second hand.

A further development in the early 1900s was the invention of the Somerville sounding gear which allowed the lead to be cast well forward of the ship. Use of this machine allowed for soundings to be carried out down to about thirty five fathoms at a speed of five knots.

> On Lake Superior [in the early 1900s] the water was very deep close inshore — too close for safe ship sounding. When a gig was stopped for a sounding — and because of the possibility of pinnacle rocks [the stops were frequent] — sometimes ten or twelve minutes would be consumed in letting the lead reach bottom and then hauling it up again. The surveyor had time to check his notes, fill his pipe, and if the day were warm, almost fall asleep. The oarsmen lazed idly. The one who worked hard was the leadsman.
>
> R.J. Fraser
> *Dominion Hydrographer*, 1948–52

During the 1920s, Canadian hydrographers explored the use of several other methods for sounding depths between 60 and 250 fathoms while the sounding ship remained underway. One of these involved the replacement of heavy deep-sea line on a Lucas sounding machine with very fine wire. This allowed for a much speedier operation of the apparatus so that the ship could maintain a speed of seven knots.

Another invention, used for discovering any shoals missed between soundings, was the submarine sentry, a device which when tossed overboard on the end of a wire would dive to a pre-determined depth. Towed behind the ship, it would eventually strike a place where the bottom rose, ringing, through another series of connecting wires, a bell onboard the ship. It would then be released by a trip wire and float to the surface for retrieval.

The harpoon sounder used a seventy-pound lead attached to a depth measuring device, recording fathoms according to the rotations of a small propeller. Attached to a wire, the sounder and lead were dropped overboard. A clapper mechanism stopped the rotation of the propeller as it hit the bottom. Upon retrieval, the depth was read off recording dials attached to the sounder itself. The Kelvin pressure tube, operated from the ship in the same fashion as the harpoon sounder, registered depth as a function of water pressure.

Often enough to prove annoying, sounding with *any* of these devices could miss some underwater hazards. The pinnacle of a rock or the masthead of a shipwreck might easily remain undetected between the repeated runs of a survey launch. Even with modern electronic equipment such hazards can pass unnoticed; an echo sounder, for instance, may not reflect the shipwreck's mast peak or rocky pinnacle unless the sounder is used directly above the obstacle.

To overcome these deficiencies the hydrographer often resorts to "sweeping" an area where he suspects undiscovered hazards exist. The sweep is performed with two boats running parallel courses some distance apart. Between the boats is suspended a submerged wire. When the wire snags on a hidden obstruction the boats move in to pinpoint its position and depth. The first wire sweep conducted by the CHS was in 1907 at Key Inlet Harbour on the northeast shore of Georgian Bay and boats attached to the survey vessel *Bayfield* were used. The Canadian Northern Ontario Railway Company intended to build a terminus there and it was decided that sweeping was the only sure method

With an accurately positioned "stretchline" to guide him, a hydrographer sounds a shoal off Pointe au Baril in Georgian Bay in the 1960s.

of determining the existence of any underwater hazards to large coal and iron ore freighters expected to use the port.†

All of these instruments and methods of depth sounding were in use by the Canadian Hydrographic Service until World War II but by the 1930s they began gradually to be supplanted by the new echo sounders, although, even today, the lowly hand lead remains the tool hydrographers use when checking minimum water depths over shoals. Also, technology has still not found a completely satisfactory replacement for the wire sweep. But, for ninety percent of a hydrographer's work investigating underwater topography, it is the echo sounder, or some variation on original echo sounding principles, to which he turns.

The invention of the echo sounder was the most significant technological breakthrough in the history of hydrography since the day Captain Cook began to base his surveys on onshore triangulation. Experiments using the speed of sound through water, as a gauge for determining underwater topography, began during World War I as the Allies sought a means of detecting enemy submarines.

In Canada, experiments were first conducted in 1915 to determine the velocity of sound in sea water. These tests, using a sound-producing oscillator, towed by the survey vessel *Cartier*, and hydrophones suspended from survey launches moored at variable distances from the ship, were successful.

But the problem of creating a machine which could both transmit and receive a sound signal, and then accurately translate that information into a depth sounding, was not solved satisfactorily until 1925 when the Royal Navy installed the first operational echo sounders on its ships. Mathematically the problem can be expressed by the simple formula $d = \frac{1}{2} vt$ where d is the depth to be determined, v is the velocity of sound through water, and t is the time

†An example of the difficulties of sweeping operations carried out by CHS was that undertaken in the approaches to Hopes Advance Bay in Ungava Bay by D'Arcy H. Charles in 1957. There, the variance between high and low tides is 13.4-metres, not far short of the 16-metre variation found in the Bay of Fundy, the largest tidal variation of any place on Earth. Tides ebb or flow four times daily creating — in this instance — a difference in water level of more than two metres every hour or thirty-three centimetres (thirteen inches) every ten minutes. With such radical changes in water level occurring in such short periods of time, Charles was forced to adjust the depth of his sweep every five minutes.

required for the sound pulse to travel from the echo sounder's transmitter to the sea bottom and back to the sounder's receiver. So, in effect, the echo sounder measures time but displays it as depth.

In Canada, the first echo sounder, made by the firm of Henry Hughes and Son, later known as Kelvin-Hughes, was installed in 1929 on the survey ship *Acadia*.

Writing in a *Canadian Surveyor* article in 1949, former Dominion Hydrographer R.J. Fraser, recalled his initial experience with these first echo sounders: "This machine was the hammer type, a compressor operating a hammer against the bottom shell of the ship. Down in the engine room and aft in the hydrographers' quarters you could always tell if the machine was operating by listening to the hammer blows. . . The operator had to wear earphones, tune in the echoes, and read off the depths as they flashed on a recording dial."

Meehan remarks that unfortunately "with a rapidly changing bottom, the true echo [as opposed to the actual sound of the hammer] was frequently lost. When this happened, the recording gear was useless until reset with a check depth by the Lucas sounding machine (kept always in readiness)." However, at the end of the field season, Captain Anderson, the chief hydrographer, was able to report that "this sonic device proved efficient [when it was working] as a depth measuring machine to a degree of precision of 98½%." And although these Hughes echo sounders were, in 1929, priced at between four and six thousand dollars, their recognized potential was such that machines were installed the next year on the survey vessels *Lillooet* and *Bayfield*.

These early models, however, were designed to be most effective in deep waters. In shallows it was difficult for the hydrophone operator to distinguish the true echo from the original pulse, as the period of time between the two was so brief. An example of the difficulties this posed was the fact that the crew of the *Bayfield*, when exploring the famous Superior Shoal — lying in the track of the deep draft vessels in the uncharted deepwater central part of Lake Superior, about forty miles from the nearest land — in 1930, had to rely finally on the use of the lead to properly locate this hazard to shipping. In the end the shoal was found to have twenty-one feet of water over its summit.

The magnetostriction transducer recording sounders, introduced to the service in 1933, were the first to record *both* shallow and deepwater depths

A lead, bent onto the wire from the drum on the Kelvin sounding machine is dropped to the bottom. The impact of the lead on the bottom and the resultant slack in the wire cuts off the machine and the length of the wire; i.e. the depth, can be read off on a graduated brass plate on the top face of the machine. The wire is then wound in by the electric motor — see the control switch on the left of the machine. A light, top right, is fitted for night operations.

with a high degree of accuracy. Automatic recording, whereby a stylus attached to the sounder's receiver recorded the soundings at the rate of four per second on specially prepared paper rolls, was installed initially in 1930 on the *Acadia* and had become standard equipment by 1932.

The magnetostriction echo sounders, which used an inboard electrical oscillator to replace the old pneumatic hammer, had advantages beside their increased accuracy in shallow waters. They eliminated "the wear and tear on the ship's hull from the continual operation of the pneumatic hammer, not to mention the constant noise and vibration": Meehan. Also, these new sounders could be easily installed on the ships' small survey launches, thereby greatly increasing their effectiveness as tools for everyday use in the field.

The transition from leadline to echo sounder was revolutionary. As Royal Navy hydrographer, Commander B.S. Dyde remarked: "With the introduction of the ultrasonic echo sounder and its automatic recorder, hydrography moved into a completely new direction".

As early as 1932, the Canadian chief hydrographer recognized that "the adoption of the echo sounding method of obtaining depth measurements . . . has increased the sounding of the deep water areas off shore not less than thirty percent." And in 1933 Henri Parizeau, officer in charge of the Victoria, British Columbia office, declared that "the most important development in hydrographic surveying is sounding by echo."

By 1935, all the major CHS ships, plus their auxiliary survey launches, were equipped with echo sounders. And, following World War II, the introduction of SONAR (sound navigation and ranging), a more sophisticated refinement of the echo sounder, completed the basic arsenal of depth sounding instruments available to the modern hydrographer.

It was complete, that is, until the CHS began in earnest to map the sea floor beneath the polar ice shelf in Canada's Arctic.

Sounding through ice presented Canadian hydrographers with a brand new challenge and they approached it with gusto. The year 1959 found them blasting holes through the ice with dynamite, then lowering a lead into the water below by means of a Lucas sounding machine winch. This procedure was expensive, messy, time-consuming, not very productive, and extremely dangerous.

In 1960 the hydrographers returned to the ice, this time to drill holes for the lead. This was neat, tidy, and safe but still, at minus twenty degrees Fahrenheit, most hydrographers were not enthusiastic about the amount of time it took to drill a hole through six feet of ice — not to mention the time required to lower and retrieve a lead from two hundred fathoms.

The real breakthrough came in 1961. It was thought by other scientific personnel working with the Polar Continental Shelf Project (PCSP)† that a modified sonic transducer, the combined transmitter and receiver of echo sounding devices, placed on top of the ice could be used to measure its thickness. This assumption proved incorrect, but hydrographers on the site discovered that if all snow were scrupulously removed and the flat surface of the transducer sealed to the flat, smooth surface of solid ice with a thin layer of motor oil, a sounding could be made through the ice to the sea floor. This method was accurate and faster than either blasting or drilling holes and, refined over the years, is still in use.

However, even more sophisticated methods were developed in the 1970s. In 1977 the spike transducer was put to use in the Arctic. This invention involved a transducer, attached to a steel spike, mounted on the back of a tracked vehicle. When the vehicle stops to take a sounding, a pneumatic apparatus drives the spike into the ice and the hydrographer, from within the relative warmth of the vehicle, records his sounding as from a launch or a ship. Then, in 1978, the spike transducer system was mounted on a helicopter.

Helicopters were first used in Arctic hydrography in 1954 by hydrographers assigned to HMCS *Labrador*; CHS acquired its first helicopter in 1957 for use on CSS *Baffin*. In 1963 helicopters were first used for sounding in open waters using a device called a "fish" towed behind. These fish are sonic transducers mounted in aerodynamically shaped hulls which are towed just below the surface of the water, usually by a high speed ship or launch. It was thought that the concept could be applied with even greater efficiency using a helicopter to tow the fish. Unfortunately, after a couple of seasons of experiments it was acknowledged by all, particularly the pilots, that the use of a fish by a helicopter

†The Polar Continental Shelf Project is part of the federal Department of Energy, Mines and Resources and provides logistic support to a wide range of research programs in the Arctic.

was too dangerous. It was too difficult for a helicopter pilot to maintain the transducer at a constant depth beneath the surface of the water; keep to a rigidly defined course determined by the hydrographer's sounding lines; maintain proper tension on the line towing the transducer; and fly the helicopter, all at the same time. But the later combination of spike transducer, attached to a pneumatic apparatus, mounted on a helicopter which then moved across the surface of the ice much like a giant grasshopper, proved a boon to the hydrographers as, obviously, many spot soundings over a wide area could now be achieved quickly.

These then are the basic tools, instruments and methods that have been and are being used by hydrographers in the field. Though over the last hundred years there have been marked changes in the hardware available and consequently the methods used, the field hydrographer's purpose remains exactly the same as in the days when Captain Cook first surveyed our waters. The hydrographer in the field is gathering information for publication of charts, *Sailing Directions*, and Tide and Current Tables.

While at sea during the survey season the hydrographer looks with envy on the eight-hour working day enjoyed by most of his shorebound contemporaries; *he* must take advantage of every hour of suitable weather to gather data, and if that means twelve hours a day in a survey launch tossing about on a stormy sea . . . so be it. And, more often than not, his day doesn't end when he climbs aboard the ship after his tour of duty in the launch; "burning the midnight oil" is not a flippant phrase to a hydrographer but the harsh reality of survey life.

Much of what he does one day determines what he must do the next, so when he disembarks from the launch the hydrographer must process the data he has collected. Today, with the increasing use of computers aboard ships of the CHS — and with computer terminals installed even on some launches — this task of data compilation is becoming less onerous. However, on smaller surveys even today, the work must still be done by hand. Even on surveys with a full range of computerized equipment, the field hydrographer who regularly works a forty-hour week at sea is the exception rather than the rule.

There's no doubt, though, that computers are easing the hydrographer's workload. But he is in little danger of becoming a button pusher only. The computer can be no better than the information that is fed into it, and the quality and reliability of the data are still in the hands of the individual hydrographer. While the electronics revolution is assisting the hydrographer, it is still his personal dedication to accuracy and thoroughness that ensures the safety of seamen, ships and cargoes.

So they're pleased to have the computers and the "black boxes" on board but they don't rely on the machines to do their job for them. The computer, especially, is, in the end, just another tool.

In the final analysis it is men not measures that count — men who check and recheck the data that their wonderful machines have delivered.

Ultimately, in the last weeks of autumn, when the waters begin to freeze and the seas swell and the winds begin to cut like steel, it is men who carry ashore the many bits of plastic, paper and printouts that will, over the winter months, be made ready to begin the process that is the making of a chart — that is the *art* of cartography.

Georgian Bay, Kingston, Ontario; Lac St. Jean, Quebec; Jervis Inlet, B.C.

Charting of Canada's inland waters is essential for transportation and recreational boating. Survey parties at Thessalon on Georgian Bay and Kingston, Ontario and Lac St. Jean, Quebec make the precise measurements and gather survey data that will be used for the development of revised up-to-date charts for these areas.

The narrow fjords of British Columbia's coastline provide recreational boaters with some of the best fishing and most spectacular scenery on the continent. The topography and bathymetry of these inlets are unique in Canada as the inlets drop from the mountain tops steeply down to depths of several hundred metres.

To provide up-to-date charts of these waters, the survey crew works from a small launch, taking soundings.

OR BEFORE
CHECK OIL LEVEL
CHECK COOLANT FOR
SIGNS OF LEAKAGE
REMOVE LOOSE DIRT
FROM COMPARTMENT
EVERY 50 HOURS
CHECK BELT TENSION
ON WATER PUMP

The Charts

A chart is one of the essentials of the navigator and the information concentrated on the surface of that sheet of paper is his guide to security. . . The chart. . . it is a work of human art, the result of information gathered from many sources and compiled with great care and accuracy. Yet few seamen, and still fewer landsmen, know anything about the work which is necessary to produce the invaluable guide to navigation.

THESE WORDS WERE SPOKEN BY R.J. FRASER, A FORMER Dominion Hydrographer, at the end of a radio broadcast in March 1934; he was quoting directly from *Shipping World*. Fifty years later, though much has changed in the chartmaking process, the essence of these remarks remains true, particularly the fact that most of us have little conception of the amount or the kind of work that is required to make a chart.

Perhaps, before getting any further into what makes a chart, it might be best to understand what a chart is. In some senses a chart is a map; and in other ways it is quite different.

A chart *is* a map to the degree that it portrays the land beneath the water the same way that a map depicts the land above; it's the difference in depiction that is significant.

A map — depending on the *kind* of map consulted — may show mountains, rivers, cities or smaller settlements, and the routes — roads or highways — connecting them. All of these landforms and indications of humankind will be visible to the mapreader's eye when he approaches them; he'll be able to

It is better to have absolutely no idea where one is, and know it, than to believe confidently that one is where one is not.

César-François Cassini de Thury
Eighteenth-century French surveyor
and topographer

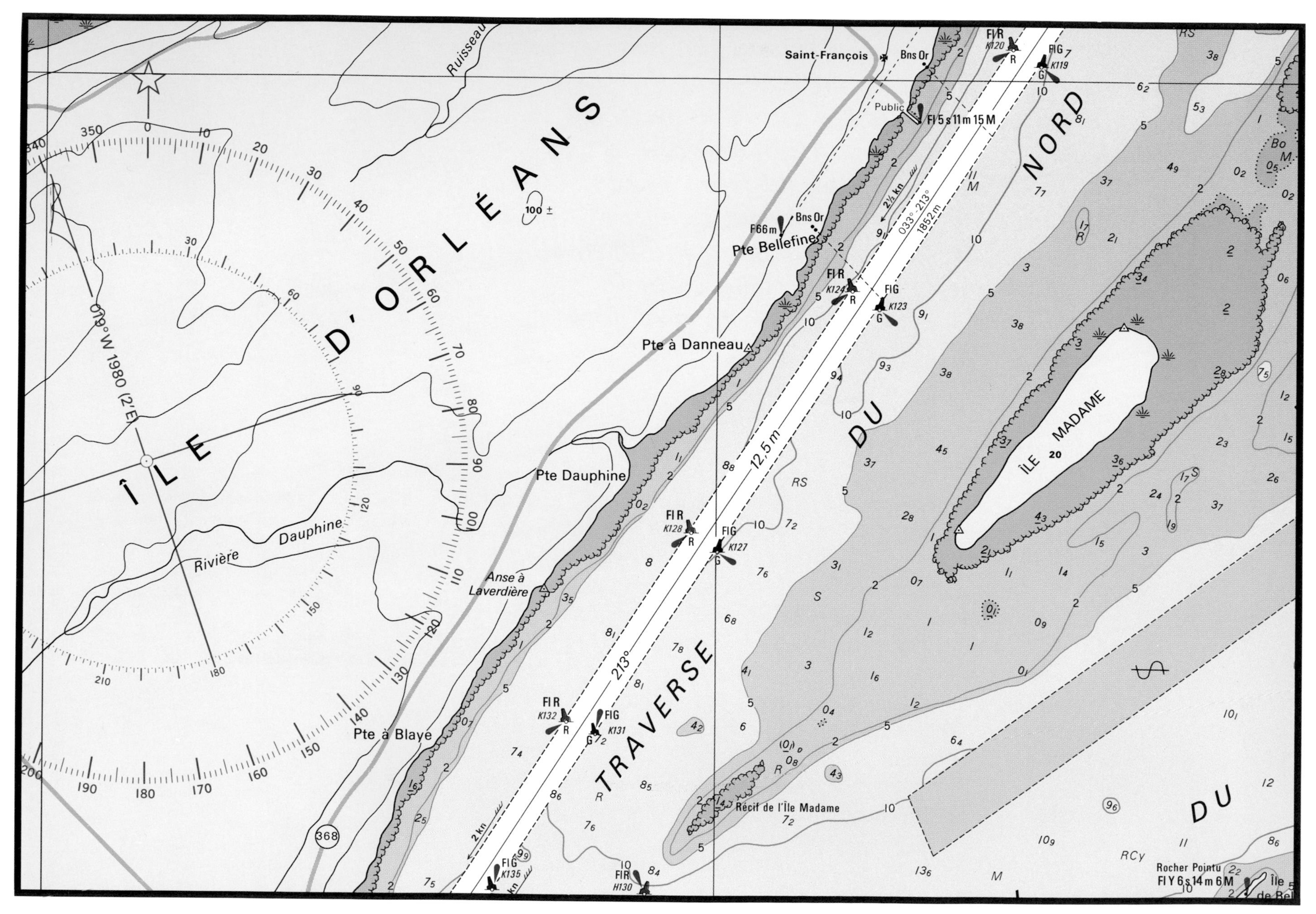

This section of a chart of the St. Lawrence shows the dredged and buoyed channel through the Traverse du Nord. The channel is maintained to a minimum dredged depth of 12.5 m. Thus no soundings are shown throughout its length. The channel runs 033°/213° and is 1852 m long. Ile Madame has a marked survey control station at each end of the island and a spoilground area lies to the south of it. There is a church spire at Saint-Francois (upper right).

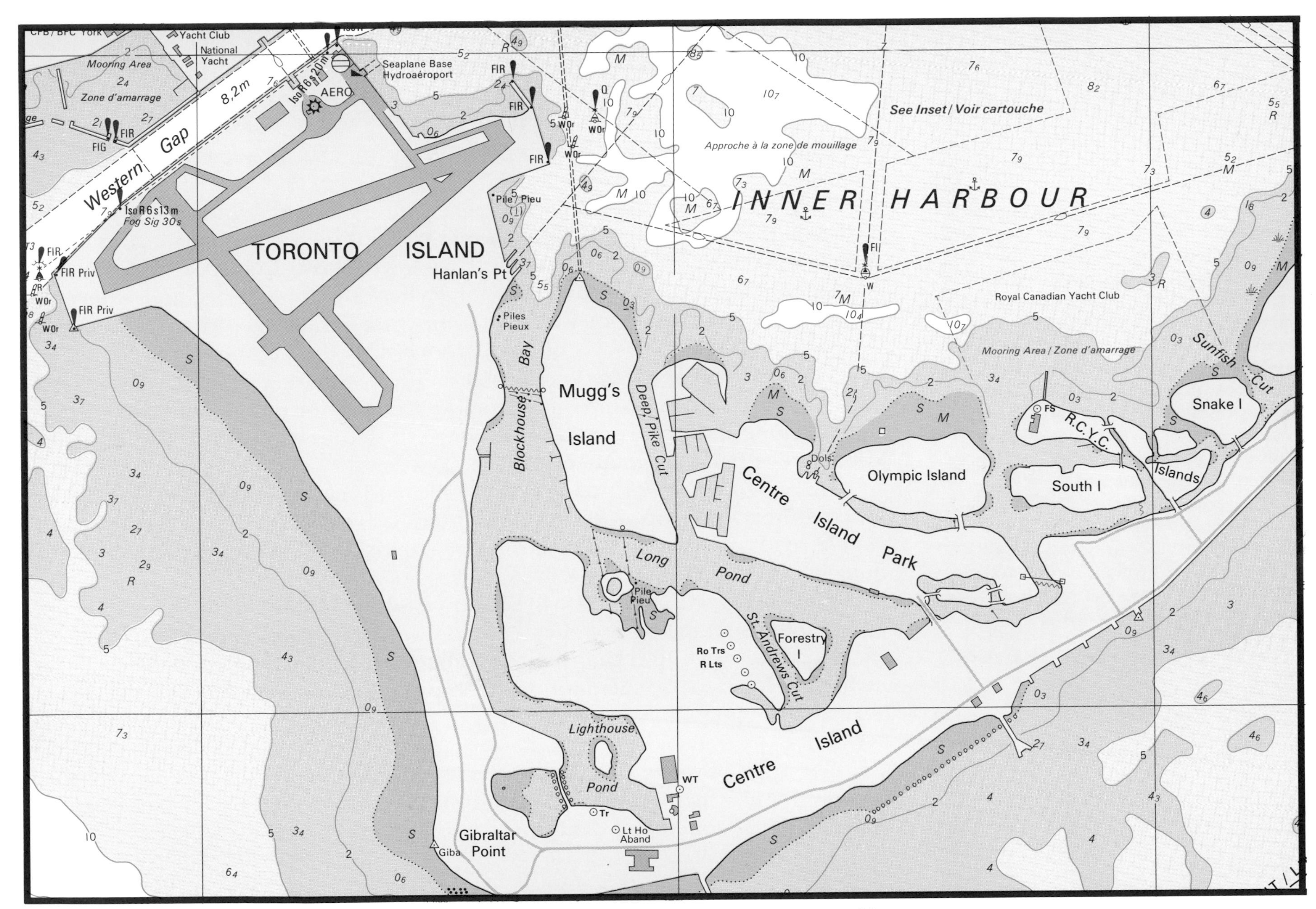

The Toronto Harbour Chart shows the anchorage in the harbour. Submarine pipelines and a submarine cable are symbolized in Blockhouse Bay.

What the chart should show

Rear Admiral Sir Francis Beaufort, for whom the Beaufort Sea was named, was Admiralty Hydrographer from 1829-55; his surveying techniques followed in the tradition established by Cook and confirmed by Vancouver. His instructions issued in 1831 to survey ship captains remain applicable today: "The first and principal object will be the hydrographic contour of the Gulf with its islands, roads, shoals and soundings . . . an accurate examination of the entrance . . . and judicious marks for passing over the bar in deepest water. . . . The heights of all headlands, isolated hills and remarkable peaks should be trigonometrically determined and inscribed on their summits on the chart — as they afford the seaman a ready means of ascertaining his distance . . . The nature of the shore, whether high cliff, low rock or flat beach, is of course inserted on every survey — but much more may be easily and usefully expressed — for instance the general elevation of the cliffs and their colour, the material of the beach, mud, sand, gravel or stones etc. . . ."

confirm his location in a particular area by the correspondence of his visible surroundings with the symbols on his map.

The same thing is not true when the same person studies a chart. The bottom of an ocean, lake or river may be metres — or thousands of metres — out of sight; the user of a chart cannot confirm the chart's accuracy by a visual check.†

Related to this is the fact that maps — at least those with which most of us are familiar, roadmaps — assume that we will travel from point *A* to point *B* in the most direct route along clearly marked roads. As a result, roadmaps confine the information they contain to where the traveller is, and where he wants to go; what lies along his route — or what lies off his route if he wanted, or was able, to vary it — is of small concern to the mapmaker.

The chartmaker must contend with a different kind of traveller. The master of a merchant ship may ship out from port *X* bound for port *Y*, only to be intercepted along the way by a radio message to detour via port *Z*, hundreds of kilometres off his original course.

He replaces the chart on his chart table and then — unlike the mapreader who cannot wheel off until the next interchange — changes course with no visible road, no signs to guide him. The "roads" and "signs" are available to him, however, on his charts.

The deepwater channels will be shown on the charts and the hazards, such as sandbanks and shoals, indicated. Contour lines indicating depths will be shown so he can avoid the shallows where his ship would touch bottom. If he happens to be a captain piloting his ship through the Beaufort Sea, the charts available to him will indicate the areas of pingo-like features (PLF) that he must avoid.

There are other ways in which charts are different from — and more informative than — maps; some of these differences will become obvious before this chapter is concluded. Not the least of these differences, to mariners who

†While he is deprived of visual confirmation of the bottom over which he's sailing, the modern navigator — commercial, and in many cases, recreational — is equipped with various kinds of electronic devices which to a large degree replace the affirmation of his eyes.

sail from nation to nation, is the fact that most of the symbolization on charts is standardized. Maps, intended largely for the consumption of nationals, are not normally standardized.

In most cases (there are exceptions) the user of a CHS chart *knows* that the symbols used correspond precisely to those used internationally by all coastal nations — some fifty or more — that belong to the International Hydrographic Organization.

That the creation of a nautical chart is a long and complicated procedure should be evident from the vast amount of information conveyed by this one piece of paper. And a useful way of approaching an understanding of the chart-making process would be to first examine, in detail, the information a chart presents to its user.

Most obviously, there is usually a shoreline. Then there is a compass rose by which the mariner can plot a course in terms of a heading or bearing. There are the sounding depths and lines of depth contours or isobaths so that a ship drawing, say, ten metres of water can avoid water that is only eight metres deep.

Shoals and rocks are shown by appropriate symbols and abbreviations, as are dozens of other hazards, aids to navigation, and conspicuous features in the water, along the shore, and on the land. The "dictionary" used by mariners to comprehend all these symbols and abbreviations is published by the Canadian Hydrographic Service. It's known simply as ***Chart 1, Symbols and Abbreviations used on Canadian Nautical Charts***. Because of the complexity of the system used to identify so many bits of information, a navigator — an inexperienced navigator, at least — without ***Chart 1***, would not be "lost" exactly, but he could well be heading for trouble, or actually in trouble. His chart probably contains the help he needs, if he but knew how to interpret it.

Also noted on the chart is the scale to which it has been drawn. In what has been called a "classic study" on the choice of scale in hydrographic charting, Vice Admiral Sir Archibald Day, Hydrographer of the Navy (1950-5), wrote that, "generally speaking, the closer that a navigator wishes to take his ship to land, whether below the keel, or to one side of the hull, the larger must be the scale of a chart." Of course, in the world of maps and charts, large scale means greater detail of a smaller area; small scale indicates less detail over a

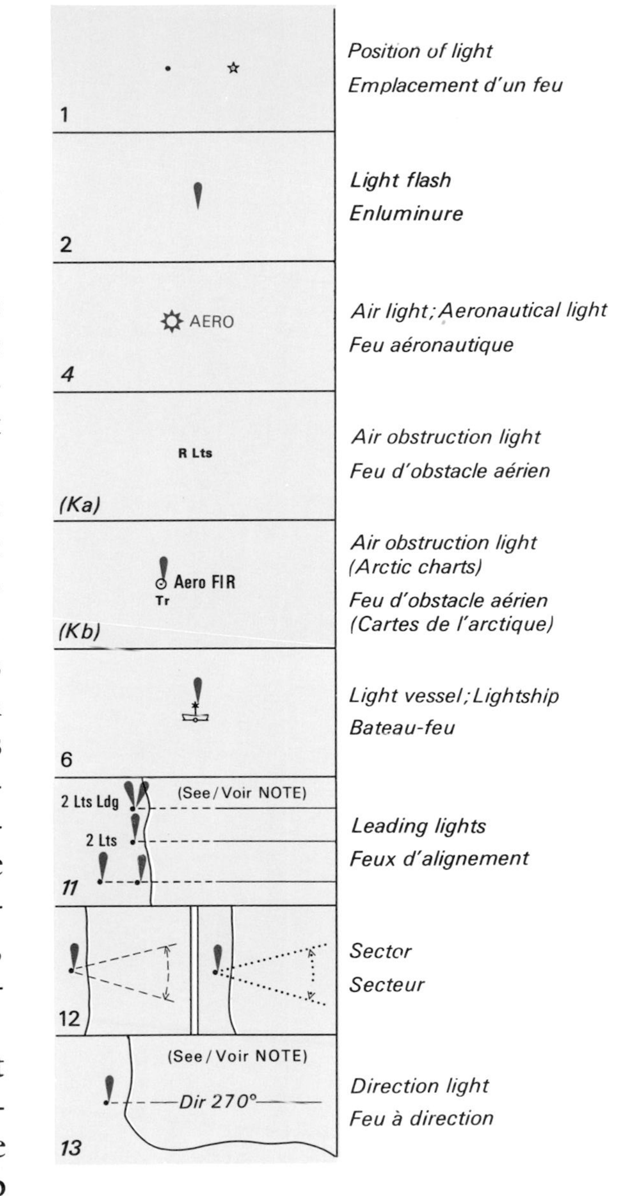

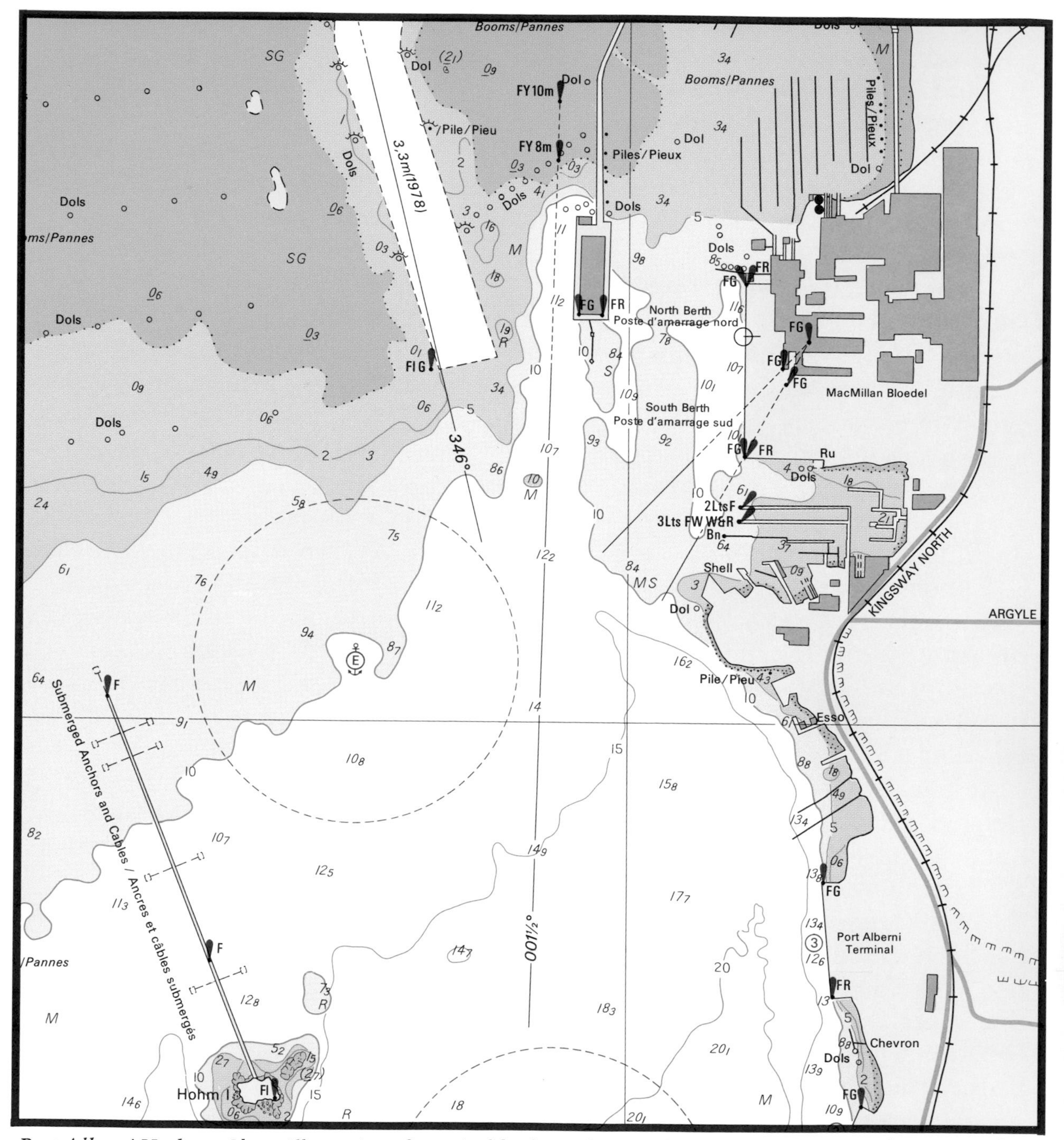

Port Alberni Harbour Chart, illustrative of a typical harbour chart. A designated anchorage is shown in mid-harbour and a floating breakwater extends from Hohm Island.

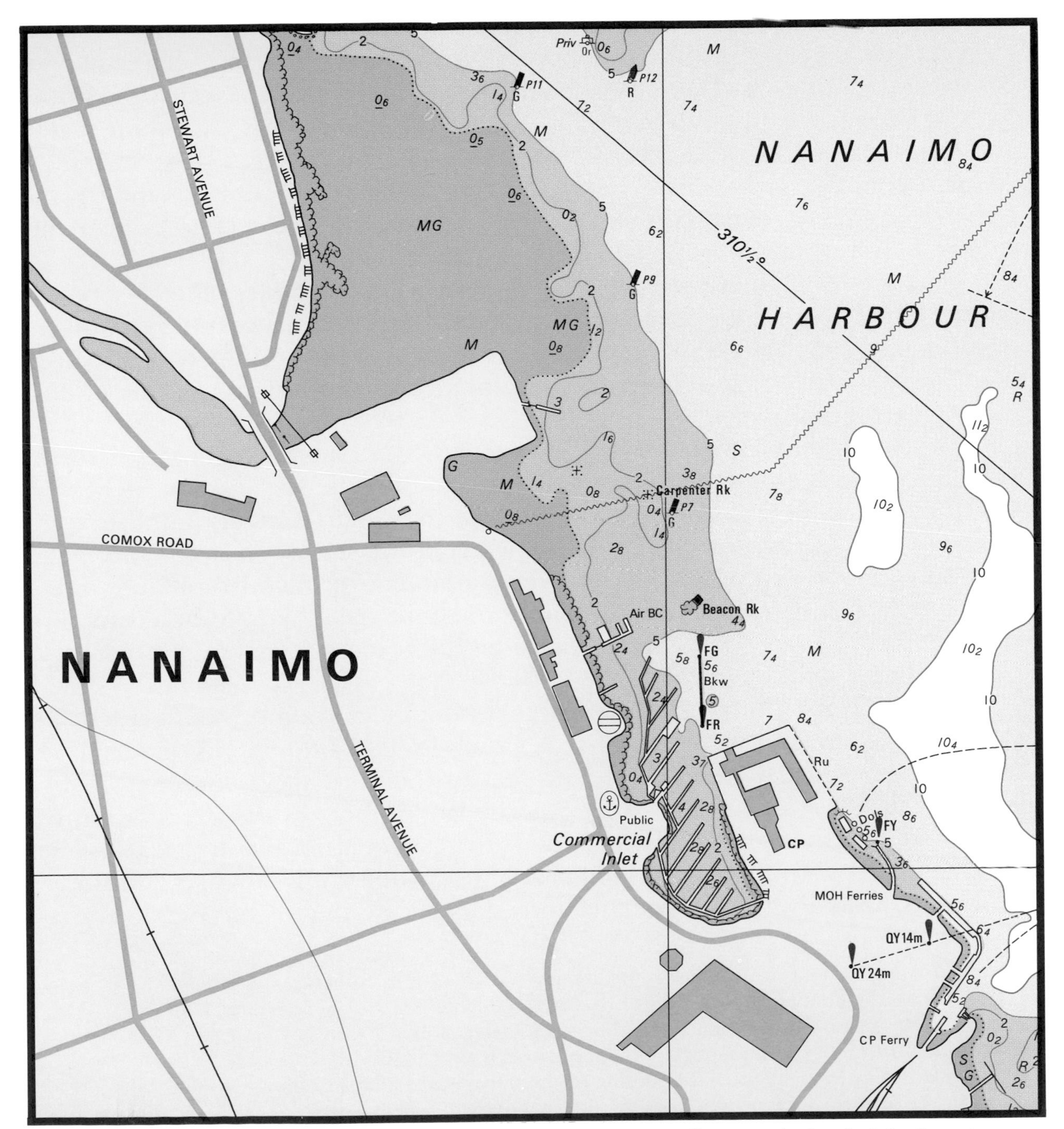

A Chart of Nanaimo shows small craft facilities. The Harbour Master's Office is at the head of the floats in Commercial Inlet; the customs office is to the north. Carpenter Rock is awash at low water and the symbolized nature of the low water areas is given.

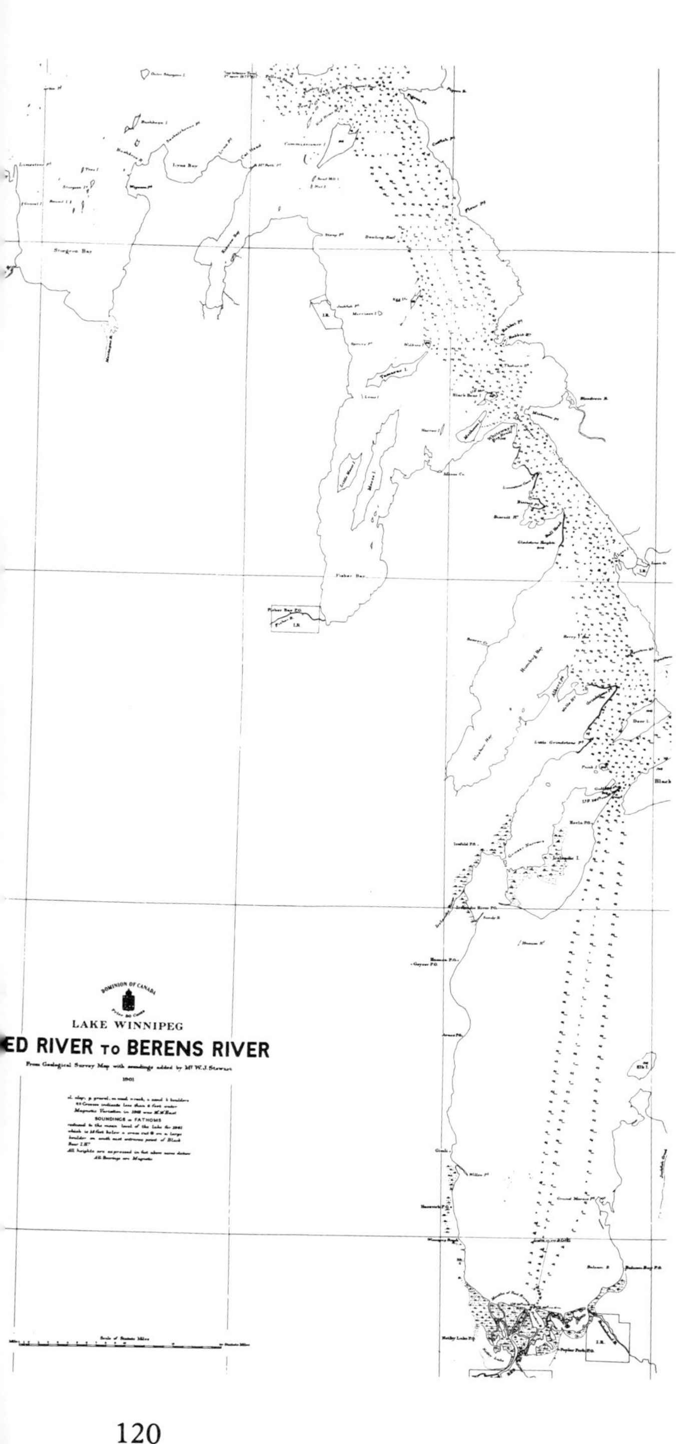

larger area.† For example, in their book, *The Maps of Canada*, Nicholson and Sebert suggest that "harbours and channels may be shown on a scale of 1:5,000 to 1:60,000; for well frequented coasts 1:80,000 is the most common scale, and there are smaller scale charts (in the area of 1:250,000) in great variety." At the other extreme are charts, created for pleasure boaters, of lakes and rivers, that employ scales of as large as 1:1,200 (or one inch on the chart equivalent to one hundred feet on the water).

That modern Canadian chart scales are shown as multiples of ten is a result largely of our joining the Monaco-based International Hydrographic Organization (IHO) in 1951, and that organization's standardization to metric measurement in its work of establishing and maintaining international standards for marine cartography.

Our first charts were heavily influenced by nineteenth century British Admiralty charts which were usually at a scale of some variation of inches or fractions of an inch to a nautical mile. The first chart actually published in Canada, from the work of William J. Stewart's 1902 survey of Lake Winnipeg, followed that tradition and used a scale of one quarter inch to a statute mile, or 1:255,594, a small scale used mainly in those days by land-based topographers. Stewart used the scale because his survey results were plotted on a topographical map.

The further influence of land surveyor/engineer training was evidenced in subsequent years by the use of scales such as 1:12,000, a fairly large scale that translated neatly into a thousand feet to the inch and sometimes corresponded to the scale employed by field hydrographers in the preparation of their own field sheets; the hydrographers' field sheet is usually drawn at a scale greater than that of their published chart. It was not until Canada's admission to the IHO, however, and the later introduction by law of metrication, that

†The scale notation in a chart's title is analogous to the f-stop in cameras: the smaller the numerical figure used to denote the opening of the lens, the larger the actual aperture. Thus, in photography, "f-stop 2" indicates a lens opened widely, while "f-stop 22" indicates a lens stopped down to an exceedingly narrow aperture. In a similar way, a scale of 1:1,200 in a chart's title indicates a large-scale rendering of the underwater and shoreline features; 1:250,000 means fewer features shown on a chart of similar physical dimensions.

Canadian chart scales approached any semblance of uniformity. And, complete uniformity is still years away — if, indeed, it is *ever* achieved.

As we have stressed repeatedly in this book, a chart's scale, and thus the amount of information it can convey to the sailor, is a function of several factors. First, the size of the body of water and the complexity of its shorelines and bottom will affect the scale of the charts that represent it. Second, while considerations of uniformity and tidiness may make the hydrographer yearn for the day when each chart will be published in the same scale as every other, the requirements of the navigator — the user of the charts — is the single principal determinant of the scale.

But, for many mariners, the scale in the title of a chart is, at least initially, merely a convenient way of indicating how much detail a chart may be expected to provide. When plotting a course, the measurement of distance is also a function of the scale of latitude. Because of the problems that arise when the spherical surface of the Earth must be represented on a piece of flat paper, apparent distances on charts and maps, particularly in extreme northern and southern latitudes, are sometimes deceptively distorted. But one minute of latitude will always equal one nautical mile, and provide a ready reference against whatever scale is being used.

The projection most commonly employed now is that invented by Mercator. An accomplished Dutch mapmaker, Gerhard Mercator published his map of the world in 1569. It bore the legend ''New and Improved Description of the Lands of the World, amended and intended for the Use of Navigators''. Those were the days of the early explorers, crossing the oceans, discovering new territories, laying claims in the names of kings and emperors to lands of riches and plenty. Small-scale charts of the time were the portolans and *routiers* such as those used by Columbus and Cabot. Rhumb lines, which gave the navigator an indication of compass direction, were included on these charts, but mariners knew well that it was impossible to plot an accurate course of any distance using *straight* lines on the *flat* surface of a chart representative of a *spherical* earth. Lines of latitude and longitude were little used in the fifteenth and early sixteenth centuries, as they presented problems similar to those of the compass's rhumb in the transition from globe to chart. What Mercator set

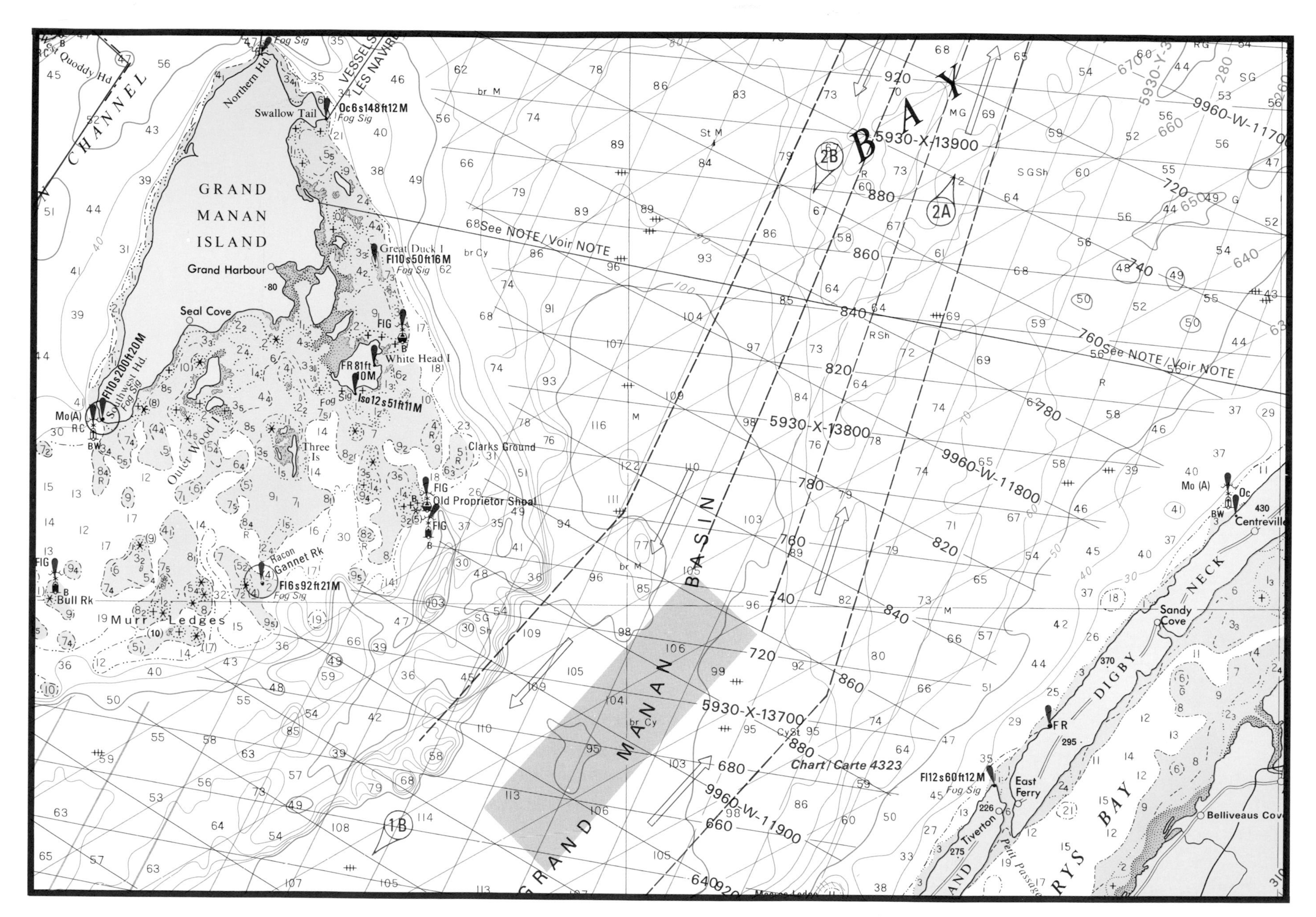

A Loran navigational lattice is overprinted in blue and brown on this chart of a portion of the Bay of Fundy. The traffic separation zones and calling in points for Vessel Traffic Management are shown in purple. Several wrecks can be seen in the centre portion. Depth contours are depicted by a mix of the new continuous line contour and the old symbolized contours.

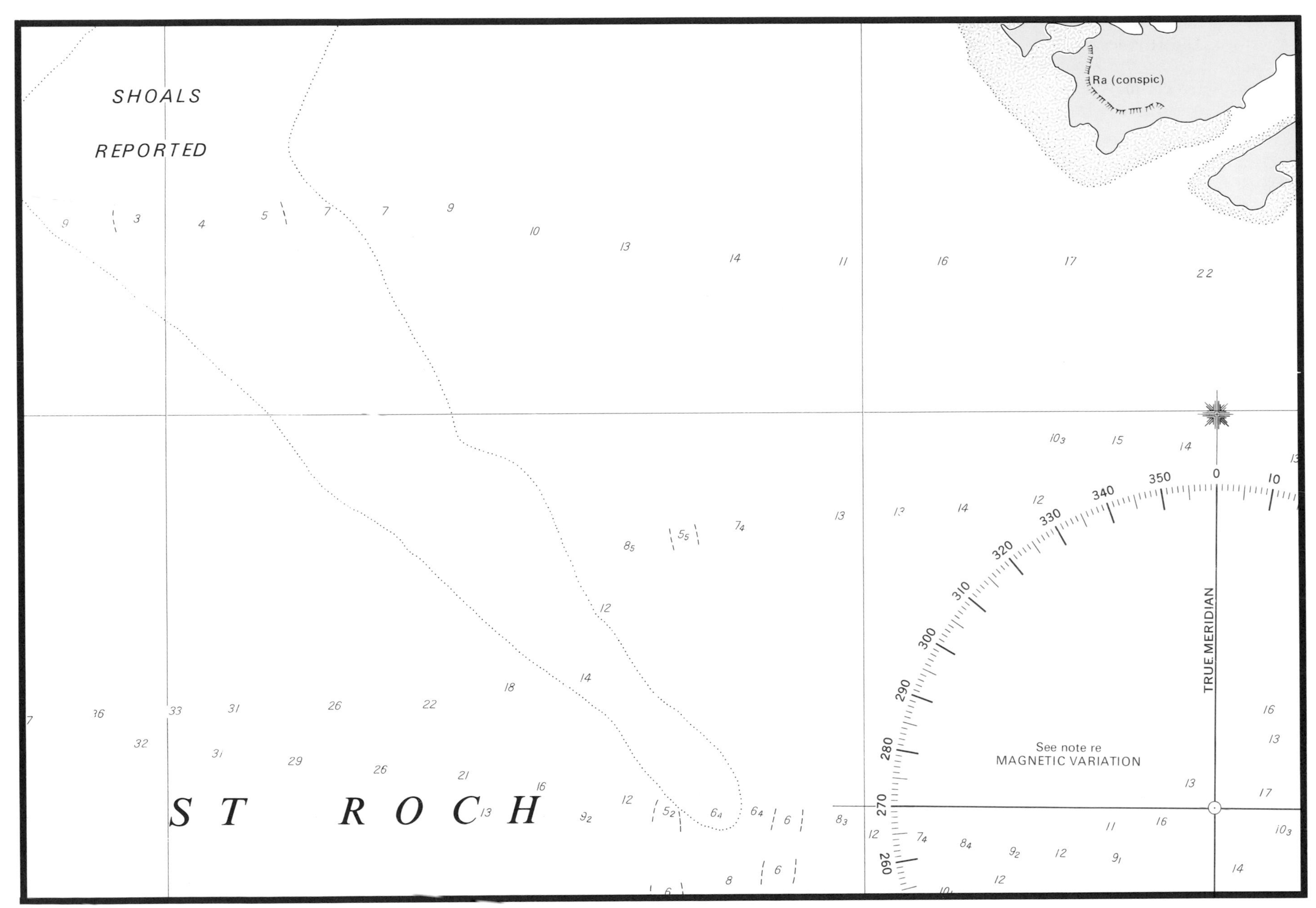

This Arctic Chart illustrates how little is known of some areas in the Arctic. The only soundings shown are from ships' track soundings whilst on passage. With no navigational aids the fact that a portion of the coastline shows up well on radar is of considerable significance in an often flat and featureless area.

In some cases, memory is the seaman's chart and old rhymes are his "Sailing Directions". . . . the Newfoundland skipper, as he steers his ship through the "tickles" (narrow straits), [sings]:

"When Joe Bett's Pint you is abreast,
Dane's Rock bears due West,
West-nor-west you must steer
'Till Brimstone Head do appear.

"The tickle's narrow, not very wide;
The deepest water's on the starboard side,
When in the harbour you is shot
Four fathoms you has got."

F.H. Peters and F.C. Goulding Smith
"Charting Perils of the Sea"
Canadian Geographical Journal
February, 1946

out to do was to provide a map upon which a navigator could plot straight line courses, over long distances, without having to constantly make adjustments to compensate for the distorting curve of the Earth. He ended up distorting, instead, the actual distances between lines of latitude and longitude closest to the two poles (and consequently the apparent size of lands and oceans in the extreme northern and southern hemisphere).

It was some time before sailors began to trust the Mercator projection, but by the beginning of the eighteenth century it had become, and remains today, the standard used for almost all charting between the latitudes of about 75° North and 75° South. Other kinds of projections are used in chartmaking, particularly in areas such as Canada's Arctic where the distortion of the Mercator projection would be too extreme for an accurate rendering of the topography. One of these projections, the polyconic, was a favourite of Canadian hydrographers of the first half of this century, especially on Great Lakes' charts, primarily because it had been used by the Americans in their surveys. Until our relatively recent adoption of IHO standards, chartmaking in Canada was, like many other of our cultural activities, a product of both British and American influences: British because of our heritage from the Admiralty; American because of our shared border, particularly along the Great Lakes. The United States' surveys, which predated any of ours, demanded some accommodation on the part of Canadian surveyors in any attempt to interface the two nations' chart schemes.

Along with the projection, the scale, the many symbols and abbreviations, the compass orientation, and the buff and blue colour outline of the shore, a chart must tell the mariner the depth of the water. Prior to 1977, actual soundings, represented by numerals indicating the depths in either fathoms, feet or metres, were printed on the chart, along with appropriate contour lines suggesting the general bathymetry of the charted area. Since that date, new Canadian charts have dispensed with many of the sounding figures. They have been replaced with other means of showing depths: contour lines and variations in colour tints. The depth contours are continuous lines connecting points on the sea floor of the same depth. The tints, ranging from white through pale to dark blue, show at a glance the relative depth of water over which the navigator sails his ship: white represents the deepest (or safest) water, dark blue the shoalest.

The new-style charts are a considerable improvement, noticed especially during night watches on the bridges of commercial ships. Then, a red, or very dim chart light illuminates the chart table as a brighter light would seriously affect the navigator's night vision every time he consulted the chart. Contour lines and tinted depth indicators show up better under the minimal illumination than sounding figures did. This move away from discrete soundings to contour lines and tints was the result of cartographers' attempts to deal with the overabundance of field sheet data provided by the sophisticated surveying technology of the years after the World War II.

Of course, not all charts are expected to serve the same public. There are roughly four main types of navigational charts produced by CHS today: for merchant shipping, for recreational boaters, for fisheries and for the Department of Defence. These different kinds of charts are all made using the same methods and production facilities. Variations are evident, however, in the scales used and in the information selected to appear on the chart. Charts intended for the use of commercial fishermen, for example, must portray a good deal more information about the nature of the sea bottom than would be required or useful to, say, the captain of an oil tanker or luxury liner; the kind of bottom below may indicate to the fisherman what species of fish may be found in the area.

Nevertheless, in the early years of the CHS, and up until about the 1950s, there was really only one kind of chart, meant for general shipping and modelled after the charts of the British. The Admiralty charts of the nineteenth century were printed in black and white only, on heavy linen paper, issued mostly in one standard size called "double elephant" (38 × 25 inches). The first charts from Boulton's Georgian Bay Survey were in fact published by the Admiralty in London. This was due partly to the fact that the Dominion government lacked the funds and expertise to engrave on copper and print its own charts, and partly to Boulton's natural reliance on Admiralty facilities. However, there were two serious drawbacks to this long-distance operation: first, an unnecessary delay from the time of a survey's completion to the actual publication of a chart was created; second, the price of that chart was much inflated. When the first,

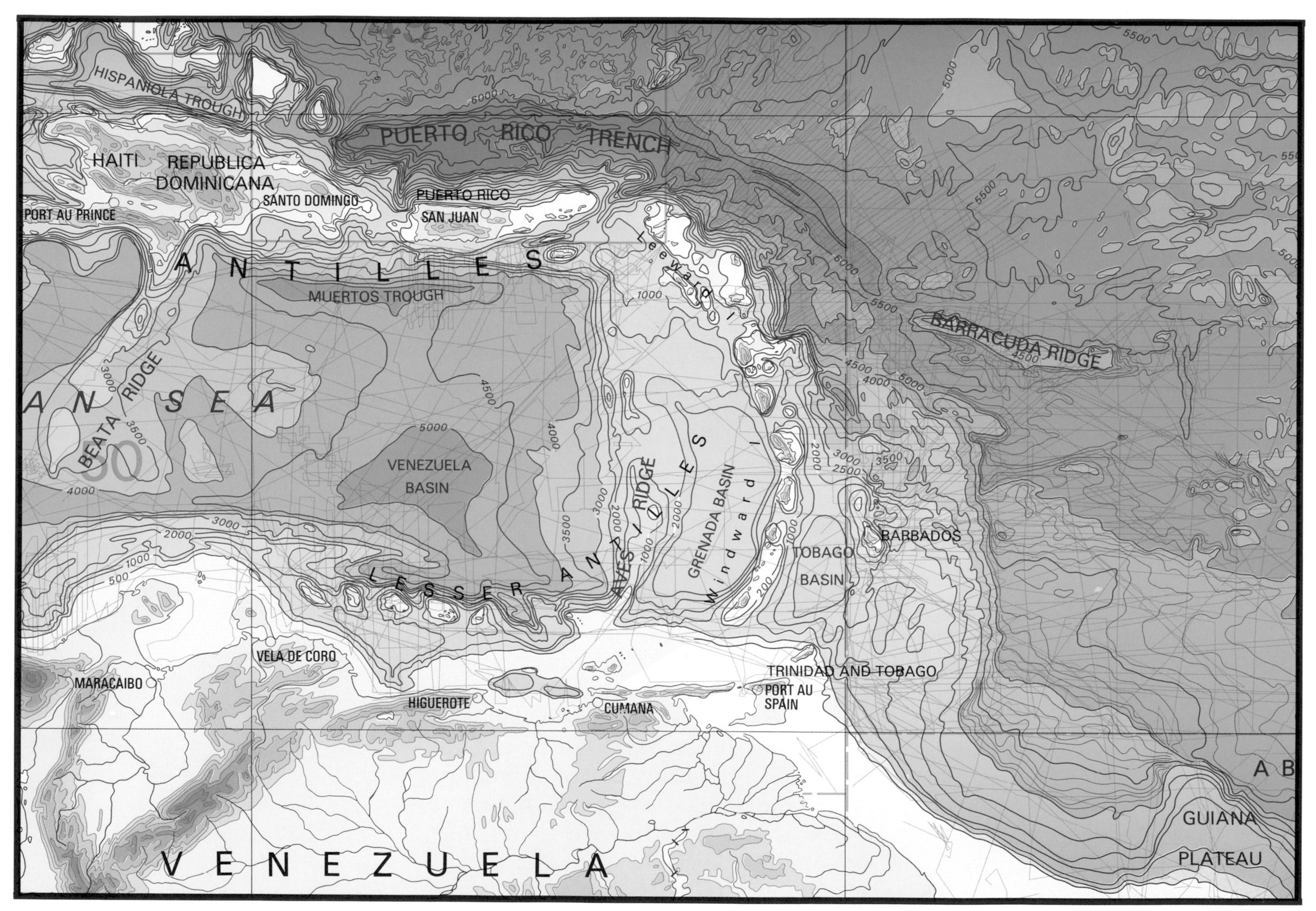

This portion of a GEBCO sheet illustrates the difference between this series and navigational charts. The deeper the blue tint the deeper the water, a reversal of the practice in most navigational charts.

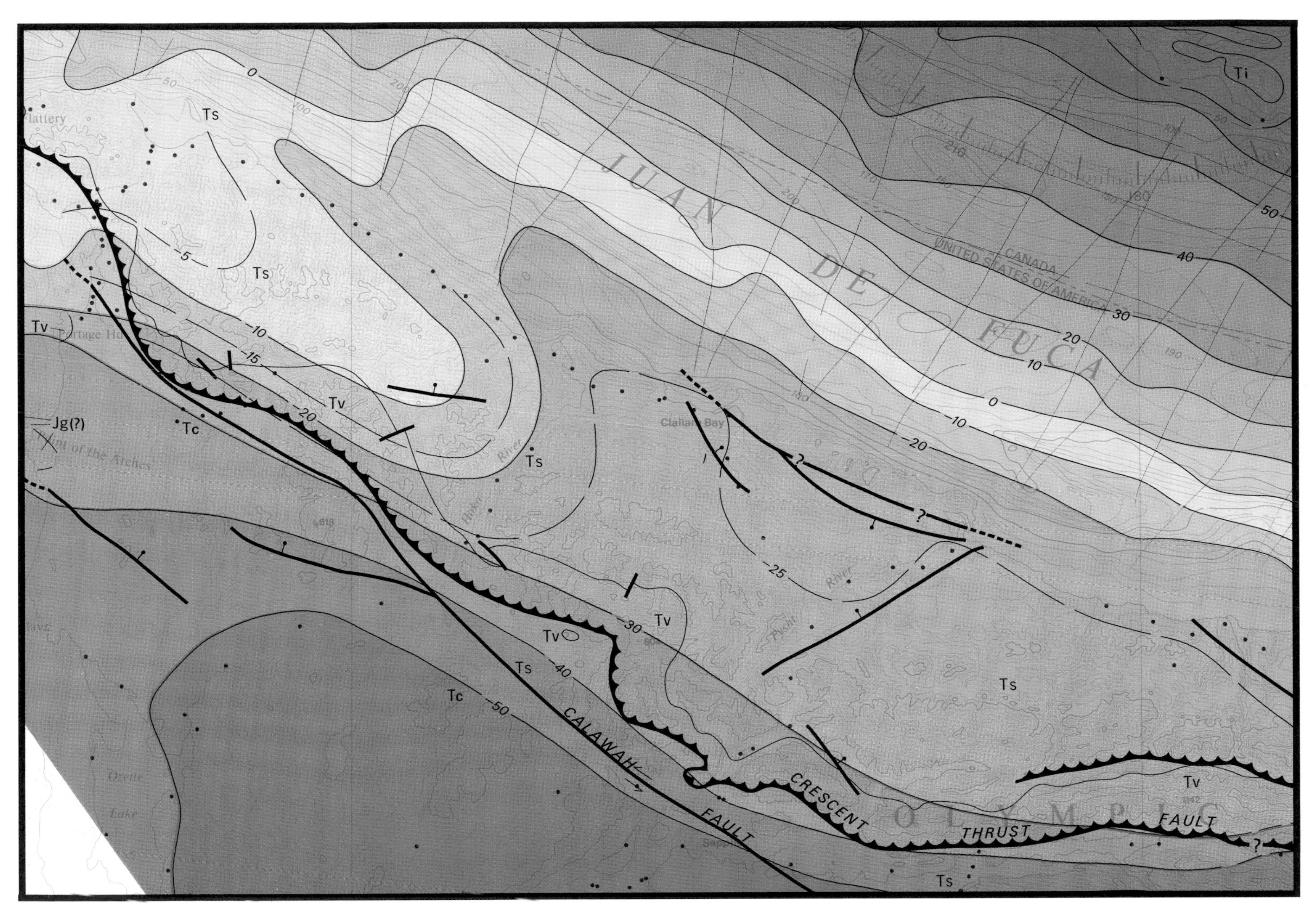

This National Resource Map of the Juan de Fuca Strait shows the magnetic anomaly of the area measured in gammas. Contours are shown as continuous lines with designators. Thrust lines are serrated and faults are shown by solid lines.

No. 906, Cabot Head to Cape Smith and Entrance to Georgian Bay was published in 1886, Boulton himself commented that as "the Admiralty published these charts at their own expense, the price was fixed at two shillings, which is very reasonable if bought in London, but when purchased in Western Ontario the price is $1.25, causing considerable dissatisfaction to purchasers, especially when accustomed at that time to free distribution of the United States charts of the American Waters."

This situation was rectified in 1903 when the chart of William J. Stewart's survey of the southern portion of Lake Winnipeg was printed in Ottawa. This was the first chart from Canadian surveys to be actually printed in Canada. It was drafted by Stewart's assistant (and successor as chief hydrographer), Captain Frederick Anderson. It was in colour, and was produced using a new photolithographic process.

Following the amalgamation in 1904 of survey facilities and personnel from Stewart's Great Lake Survey and from the Department of Public Works, the CHS began to issue charts for all of Canada's major ports. The first of these was the 1905 *Montreal Harbour, Longue Pointe to Varrennes*, printed as a colour photolithograph and drawn on the large scale, of one inch to 1,000 feet.

In 1906 a group of hydrographic surveyors were assigned on a fulltime basis to the drafting room at CHS headquarters in Ottawa. This move would eventually have a profound effect on the service, as it began the separation of field hydrographers, used to drafting their own charts, from the cartographic function which over the ensuing years would become increasingly specialized and isolated from the "field".

The first Canadian-engraved charts (engraving on a copper plate providing a finer, more accurate resolution than the early photolithographic methods) in colour, were printed in Canada in 1909. Two of these were for the Saint Lawrence River, one was of Prince Rupert Harbour in British Columbia, and one was of Georgian Bay. Three of the four were available from the "Hydrographic Survey Office" or its agents, for fifteen cents a copy. Previously, charts had been obtainable only from the federal ministry (at that time, the Department of Marine and Fisheries) in Ottawa. These 1909 charts thus marked the beginning of organized chart distribution by the CHS.

At this time, though hydrographers were busily surveying waters all across Canada, chart production was relatively limited and the nation still relied to great extent on the nineteenth century work of the Admiralty surveyors. In 1913, as war approached and the British could not handle all of Canada's requests for reprints of these Admiralty charts, the CHS started reprinting many of them by photolithography. Then, in 1914, the Admiralty shipped fifty-six copper plates for charts of the Great Lakes and the Saint Lawrence River to Canada with the intention that the CHS make the necessary corrections using information gained from resurveys, and take over the printing of the charts on a permanent basis.

During World War I, hydrographic field operations were restricted, the ships having been requisitioned by the navy, and the overall CHS budget curtailed due to the war effort. But the drafting room, using some of the personnel usually employed in the field (though many had joined the military), and encouraged by the government to produce as many strategic charts as possible, was the busiest it had ever been, producing as many as fifteen charts in one year. It was during the war that the first Canadian-made charts of the Atlantic coast (the surveys having been completed using launches and other small boats) were published. Perhaps the most important of these was that of Halifax Harbour. However, due to wartime intelligence restrictions (engendered in some measure by the suspicions aroused by the famous Halifax Harbour explosion of 1917), it was probably no coincidence that this chart was not officially published until 13 November 1918 — two days after the armistice.

In 1921 a small staff of copper plate engravers was transferred from the Dominion Printing Bureau to the hydrographic survey, in a move intended, as suggested by Meehan, "to save commuting time between buildings in Ottawa and for closer drafting-engineer relations." Also that year a new catalogue of charts was issued listing one hundred and thirty Canadian charts available to the public, an impressive number considering the limitations and restrictions under which the CHS had been working, and considering that the organization was less than forty years old.

Captain Anderson took over from William J. Stewart as chief hydrographer in 1925. Under his direction the service began to modernize and expand its

Protecting the eggs and finding the subs

During World War II CHS surveyors produced special confidential charts for defence purposes. Perhaps the most curious of these "chartlets" was the series of "egg route" charts. Compiled from studies of air and water temperatures in the Gulf and River Saint Lawrence, these sheets showed the areas of coldest water; where a ship carrying eggs — or any perishable foodstuff — could keep its cargo fresh without the need of costly refrigeration equipment. But water temperature also determined where submarines could cruise in relative safety from the probing pulses of sonar. So the egg route charts served two purposes — protecting perishable goods and providing hints on where to search for enemy submarines.

operations, mainly through experimentation with and the subsequent installation of such technological wonders as the gyro compass and the echo sounder. By 1928 the service was sufficiently confident of its chart construction division that it exhibited several Canadian charts in London, England, at the new Science Museum where they received "special mention".

As of 1930, chart construction division was issuing up to fifty editions a year, along with making 8,000 corrections to existing copper plates. Because of the expansion of the catalogue and confusion due to a lack of uniformity in the scales and projections used, it was decided to produce all new charts on the Mercator projection using scales "standardized at one inch, 0.5 inches, 0.3 inches and 0.1 inch to a nautical mile, in order to facilitate the use of a number of such charts in series": Report of the chief hydrographer. Also in 1930, the price of the 278 Canadian made charts, now listed in the catalogue, increased to fifty cents — still a bargain when compared to Admiralty and United States' prices which were at least double that figure.

And then, of course, the very next year, the effects of the depression put a stranglehold on everything, including CHS. Chart distribution decreased by twenty-five percent between 1931 and 1933. Paradoxically, this time also saw the highest levels yet of productivity in the chart construction division, a result, wrote G.L. Crichton, chief of the division, of his staff's increased "amount of assistance rendered the field parties in the completion of their fair sheets". This "assistance", given by the cartographers to the field hydrographers, greatly speeded up the chartmaking process and, like the institution in 1906 of a separate drafting office, continued the trend towards the separation of the two functions.

In 1934 chart distribution began to rise. By 1935, levels were twenty percent above the low rate of 1933. Also, authority was granted that year for the purchase of an offset press which would allow for faster, cheaper, and still accurate printing of charts. With this machine the service began to move away from the copper engraving process which had been the standard since the time of Mercator in the sixteenth century. At the end of 1937 only four charts were published from engraved copper plates. In 1947 the last of the copper engravers retired and the process was never again used in the CHS†.

†It was a moment of regret and a continuing matter of nostalgia when copper engraving dis-

By 1937 the Canadian Hydrographic Service was more than fifty years old and, although progress had been slowed by World War I and the depression, the expertise of the cartographers and the field men was unquestioned. Canadian charts accordingly had developed a reputation for accuracy second to none. In his annual report of 1937, Henri Parizeau, head of the Pacific coast office, wrote with pride that:

> many years ago boats on this coast used to go into bays and inlets without a chart, but as the development of the country increased, insurance companies refused to allow those ships to travel under these conditions and it is now such that no Coastal Company will allow any of their ships to go into any new development unless the Hydrographic Service had a survey of the locality.

(Even today, insurance company rates for ships operating in areas, such as some of those in the Arctic, not fully charted, are *very* expensive.) Parizeau went on to remark also on the benefits of Canadian charts to the lumber, fishing, and tourist industries in British Columbia and then closed off with the statement that the "Courts in British Columbia have placed the value of Canadian Charts at the same level as those of the British Admiralty and in many cases above the Admiralty Charts themselves." For a service steeped in the traditions of the legendary British hydrographers, there could be no higher praise.

In 1938 a recommendation of great import for the future work of the service was made by the chief hydrographer. This called for the production of "yachtsmen's charts", the small craft charts of today. The annual report stated that:

> these charts would cover the many sheltered, but often intricate, water routes of lakes and rivers available to yachts and motor cruisers. On account of the vigorous steps taken to develop our tourist industry, and to the growing popularity of watercraft for holidaying and recreation purposes, the demand for this type of chart is continually growing.

appeared. It was a craft — some would say an art — and its practitioners were valued contributors to the hydrographers' charts. One cannot challenge the speed and efficiency of the newer methods of reproduction. Still, hand scribing *in reverse* on a large copper plate was a memorable skill no longer taught or practised.

The future of the world was at stake

The reasons why a hydrographer must survey or resurvey a stretch of water are varied, but underlying each need is always the question of safety — the safety of lives, ships, cargoes. Perhaps never before was the question of safety so intimately involved as when four Canadian hydrographers journeyed to Newfoundland in June 1941 to resurvey the anchorage at Mortier Bay of the island's south coast. The harbour had not been surveyed since 1876 when the Admiralty had charted it with a thoroughness sufficient to the needs of the day. Though they didn't know the precise reason for their assignment, the Canadians were told it was urgent. The harbour had been tentatively chosen as a site for a meeting of British Prime Minister Winston Churchill and President Franklin Roosevelt of the United States, and the Admiralty was concerned that the anchorage was not adequate to the requirements of the two leaders' warships and their personal safety. The Canadians began their survey on 7 June and in record time produced a new chart of the harbour. It was never used for its intended purpose — Churchill and Roosevelt rendezvoused at another south coast port — but the chart did achieve a place in Canadian hydrographic history; it was the first chart of Newfoundland produced by the service, and at a time when the island was not yet a province.

At the time the recommendation fell victim to the outbreak of war. Within a few years of the war's end — by the early 1950s — the mushrooming interest in small boat cruising had again alerted the CHS to the need for charts aimed at the recreational sailor. Yet again, other events intruded and postponed their production; in this case it was mainly the charting requirements for the new Saint Lawrence Seaway that intervened. It was not until the 1960s that CHS returned again, this time successfully, to satisfy the need for small craft charts.

A sixteen-foot aluminum fishing boat powered by an outboard motor is not a Great Lakes' freighter; and a thirty-two-foot cabin cruiser or a similar size yawl is not an oil tanker. The needs of recreational boaters differ greatly from those of commercial shippers. For one thing, small boats tend, generally, to sail close to shore — it's safer to do so; the recreational sailor therefore needs much more detail, especially inshore soundings, than the commercial navigator will require. Then, too, coastal cruising is a matter more of length than breadth — it requires long narrow charts rather than sheets that are more or less square.

In 1964 the first set of accordian-pleated strip charts of part of the east coast of Georgian Bay were published. Since then CHS has continued, almost uninterruptedly, its small craft chart production; in late 1982 thirty charts were available containing eighty-one strips. For six other lakes, charts were available in folded format and one book was published.

Certainly, as the chief hydrographer's report of 1938 indicated, recreational boating had been on the rise for many years. Even so, it is difficult to escape the conclusion that much of the impetus for the sport's phenomenal growth during the 1960s through 1980s has been due to the availability of accurate and suitable charts. Indeed, so spectacular has been the increase in demand for small craft charts that today more than sixty percent of CHS's chart sales are of the small boat variety.

After World War II the chief hydrographer reported that, because of the military effort, Canada had made "a considerable advance in the total area of coastal waters charted" with "sets of navigation charts greatly improved over the pre-war issues." Yet it was also true that the peacetime requirements of commercial shipping, the heavy demand for "tourist charts", the impending charting of Canada's Arctic (which would begin in earnest in 1949), and the large backlog of corrections to be made to existing charts would place an unprecedented burden on the chart construction division. With that in mind, after 1946, an effort was made to recruit more staff (principally in the form of student assistants), chart numbering was reorganized in the CHS catalogue to conform to a more efficient geographical district system, and the office's personnel was divided into two groups. One was for chart compilation, that is, the gathering together of data from a number of field sheets. The other was called chart drafting, the actual drawing of the chart. It should be noted here that until this time it had been taken for granted that field hydrographers often took part in the compilation process. The 1946 reorganization of chart construction was another step in the separation of "field men" from cartographers.

The final major breakthrough in the history of chartmaking in Canada, that is, until the introduction in the 1970s of computer-assisted cartography, was the publication in 1953 of *Chart 4368, St. Ann's Harbour*, the first in the world to be produced by negative engraving on plastic. This procedure remains in use today, essentially unchanged. The hydrographic data are scribed (or engraved) in negative form on an emulsion-coated polyester carrier, rather than drafted on paper. The advantage of negative scribing is that no intermediate photographic steps are required to produce printing plates as the originals themselves are negatives.

Also in the 1950s, there occurred two events which led to an incorporation of both British and American charts into the Canadian system.

First, in 1954, the CHS began to reproduce a number of Admiralty charts of Canadian waters. The British were selling few of these charts and wanted to cancel them; CHS asked permission to reproduce them, as many of the charts proposed for cancellation were of areas not then adequately covered by the CHS's own surveys. The Admiralty turned over reproduction material and for ten years the CHS published as many as ninety-four of the former British charts with appropriate changes in titles and notes.

The second event occurred in 1963 when Canada assumed from the United States responsibility for supplying the DEW line stations in the Arctic. When the line was built, charts for the western Arctic had been assigned to the United States hydrographic office; in 1963 the CHS took them over and incorporated them into the Canadian system.

Since 1977, the responsibility for cartography has largely shifted from CHS

Cartographer makes changes on a scribecoat.

headquarters in Ottawa, and now much of the work is performed in the four regional offices: at Dartmouth, Nova Scotia; Quebec City; Burlington, Ontario; and Patricia Bay, British Columbia. (The Pacific Region had, since the 1930s, maintained some of its own cartographic personnel.) This total separation of field hydrographers from cartographers was the natural result of an evolutionary process begun with the establishment of the first separate drafting office in 1906.

Another responsibility shared by headquarters and the regions is that of chart scheming which is in reality where the making of a chart actually begins. Chart scheming is simply the careful planning of where and how and when the CHS will go about charting any section of coast. Requests for surveys and resurveys are received at headquarters and the regions, from such sources as boards of trade, industry, shipping, local governments, and other federal departments. After all the requests have been evaluated and charting priorities established, the service's current and anticipated resources are calculated. An attempt is made each year to plan CHS field and cartographic activities over periods of, usually, five years.

A chart scheme is, in the words of one recent CHS study:

> a small scale chart or plan covering a convenient stretch of the coast in which are displayed the limits and scales of all the charts required to permit safe navigation. . . The chart scheme enables the hydrographer to plan the limits of his surveys and is the first information required by the cartographer before he can begin to construct a chart.

This is a broad statement of the purpose of chart scheming and in its generality it conceals the subtleties of the process. To take but one example: in a passage up the Saint Lawrence River from the Atlantic, a ship's navigator will require a number of charts; no single chart could be printed in large enough scale to indicate adequately for the river's full length each shoal and other hazard to navigation. So, to start with, the hydrographer, in planning to survey the river, knows he must produce not one but several charts.

This consideration leads to the questions, "How many?" and "Where does one join to the next?" When he's answered those two questions to his own satisfaction the hydrographer is confronted with yet another question, that of scale. Since no part of the river is identical to another in either the number or the position of its navigational hazards, it follows that some charts — those showing more dangers or narrower channels or like considerations — must be larger scale than the charts for areas of relatively few hazards.

But, on the bridge of a ship, shifting from a small-scale chart of an area of open water to a large scale chart in a hazardous zone is, in itself, hazardous; it takes time to switch the sheets on a chart table — and to switch the navigator's mindset as well. Consideration of these problems forces the hydrographer to go back to the earlier question, "Where should the charts join one to another?" Now the question becomes, "By *how much* should adjoining charts *overlap* to provide the navigator with sufficient time to switch from one to the next?"

To put all of the above into context, consider the 200-kilometre stretch of the Saint Lawrence from Tadoussac upstream to Quebec City. At present this part of the river requires six different charts at six different scales. Hydrographers have been concerned about both the number of charts and the multiplicity of scales. Recently they've produced a chart scheme that reduces the number of charts required for the Tadoussac-Quebec City run to three, all to the same scale. It is a vast improvement over the previous scheme in both simplicity and elegance.

After the chart schemes have been established, the next step in the process is, of course, the field hydrographer's collection of data, and his eventual transfer of that data to the field or fair sheet, which appears as a very large scale chart of the area surveyed. (In some cases, the survey that is based on the chart scheme reveals unexpected anomalies that require changes and modifications in the chart scheme itself. The whole business of chart scheming — like so much of hydrography itself — is controlled by feedback, the constant correction of data by a constant assessment of the data's accuracy.)

At the end of the summer surveying season the transparent mylar field sheet is taken ashore, checked and rechecked by the hydrographer-in-charge of the survey, the assistant regional hydrographer, a senior hydrographer from another field party, again by the hydrographer-in-charge, and, once again by the assistant regional hydrographer before receiving the signature of the regional hydrographer.

I perceive that you have made yourself acquainted with the geological characteristics of the shores of the St. Lawrence — why not give them in the chart? . . . the leading features of the ranges of hills and the line of the coast and also the nature of the sand, gravel or shingle on the beach. . . . I think a chart should always distinguish their nature — with a separate outline for each of the above mentioned denominations.

Letter, 7 January 1830
Admiral Sir Francis Beaufort
Admiralty Hydrographer
to Commander Henry Wolsey Bayfield
Saint Lawrence Survey, Quebec City

Charts today are compiled from the data of several field surveys. Compilation begins with the preparation of a ''projection'', usually Mercator, which portrays the selected parallels of latitude and meridians of longitude at the scale of the chart being produced. The gathering together of this information is the beginning of what is known as data compilation (as opposed to data collection).

Then, using large cameras, the field sheets are reduced to the scale of the chart projection. These cameras are designed to ensure that the reduction is accurate, that no distortion occurs and as much of the detail as possible is retained. The film positives thus obtained are then cut and mosaicked onto the projection, the plotted control points and intersections of latitude and longitude on the source documents being matched with those on the projection. The result is the compilation mosaic.

Next comes the most important phase in the production of a nautical chart. This is the selection of the information that will appear on the final product. On a ''blueline'' (a blue image of the compilation mosaic photographed onto scribing material), the cartographer prepares the ''compilation copy'' by selecting the information to be charted (soundings, depth contours and topography), then manually scribing it. When the compilation copy has been completed and checked, it is ready to serve as the cartographer's basic source copy. It is in turn bluelined onto scribecoat, where the cartographer does the final drafting of soundings and linework.

Place names, symbols and abbreviations are stripped onto the names overlay. Aids to navigation, lights, buoys and fog signals are added as well as navigational aids in the form of symbols indicating the position of conspicuous smokestacks, the spire of a church or a cathedral, radio masts, towers — any shore-based landmark that will help the mariner position his ship. Cultural features such as roads, streets, buildings and railways are added to complete the presentation. Colour negatives are then prepared for printing the land tint, the gradient tints of the bathymetry and to highlight the aids to navigation shown on the chart.

At each step, the information and the quality of its reproduction are checked by quality control cartographers; no chart is approved until its final colour proof is approved by the Dominion Hydrographer himself. Ultimately, twelve to fifteen months after the field sheets were originally compiled, a chart is printed.

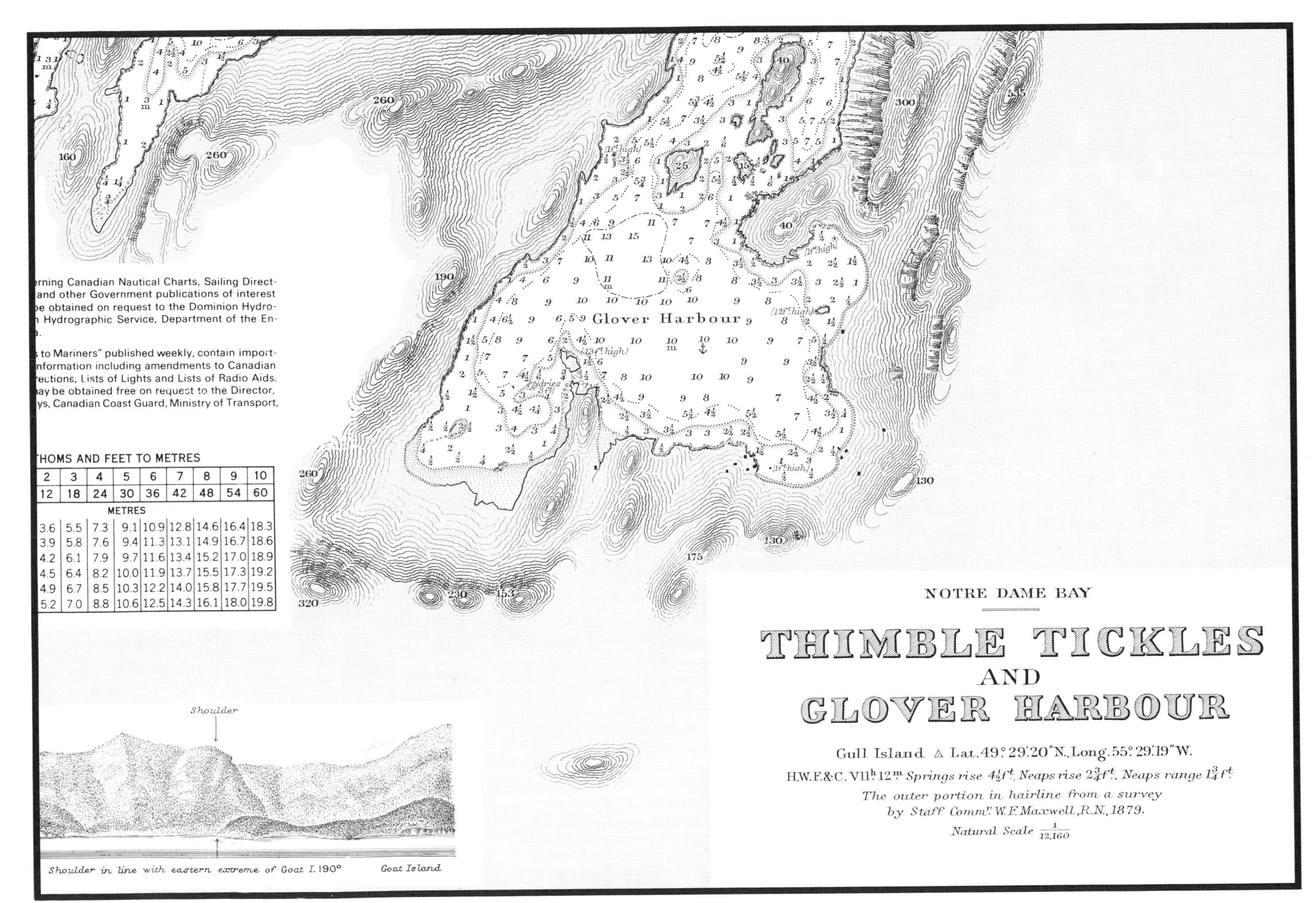

This is an example of a copper engraved chart with height contours shown by hachuring. Depth contours are shown by symbolized contour lines. The bottom left shows a view of the approaches to the harbour from seaward.

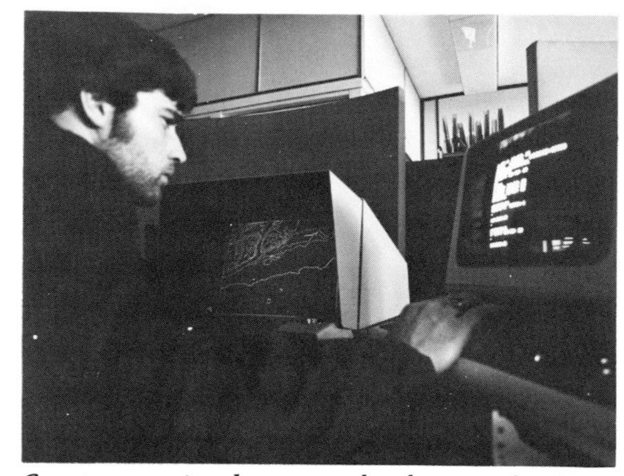

Computer assisted cartography shortens the time required to produce a chart. Here a cartographer works with electronic digitizing equipment.

In 1967, the Canadian Hydrographic Service began developing a computer-assisted chart production system. High-quality drawing systems have been developed to draw on film the various projections and borders used on Canadian Nautical Charts, Loran-C and Decca lattice overlays, and digitized chart data (shoreline, soundings, topography and depth contours). All film drawing is done by computer-controlled precision drafting tables mounted with electronically controlled optical light projectors.

Digitizing stations have been developed to convert graphical information into digital form suitable for processing by a computer. The digitizing stations have three major components: a digitizing table with a manually operated cursor, a computer, and a disc for data storage and retrieval.

More recently, CHS has developed an interactive editing station called GOMADS (Graphic-On-line Manipulation And Display System). It is used for viewing and manipulating digital chart data so that errors and omissions which have occurred during digitization can be corrected. The data is viewed on a Cathode Ray Tube (CRT) screen and can be altered by the operation of a combination of cursor-joystick on the CRT and an accurate pointer on a digitizing table. Line, point, numerical and name data can be added, deleted, altered or moved. The accurate pointer is used to allow precise positioning for additions and alterations. GOMADS is recognized as one of the premier editing systems in the world.

After a chart is printed, the up-to-date digital data are stored on tape in the CHS Digital Chart Library where they can be retrieved quickly and updated digitally to produce a New Edition of the chart. This information can also be used to produce a new chart at a different scale, perhaps on a different projection, or to produce an overlapping chart at the same scale. The first Canadian chart to be compiled and taken through all the stages of automated drafting (No. 3457, *Nanaimo Harbour and Departure Bay*) was not produced until 1979. Considerable development effort was required to reach that stage and because of the large capital investment and the training required, it will be a long time before all charts will be produced by means of the new technology. To this point, the drafting phase of chart production has been automated to some extent, but the lack of digital source data means that it will be many years before computer-assisted techniques can be applied to the more complex chart compilation process.

Now, field data are being collected in digitized form and processed by onboard computers. That is, the soundings and their positions are stored in such a way (either on magnetic tape or on disc) that they can be understood by data processing hardware used in the compilation process. The digital information is still accompanied in many cases by field sheets rendered by hand. Yet the automated production of charts is at least theoretically possible and it's expected that, eventually, computer-assisted cartography will supplant more traditional methods altogether. It remains only a matter of time until the appropriate software and hardware for the complete manipulation of field data, in digital form, are developed and installed in the hydrographic offices.

Charts are continually updated, either to correct mistakes or to incorporate new data revealed by resurveys of an area and changes in navigational aids. Most charts are amended before distribution by hand in pen and ink or rubber stamp, or, when the changes are more extensive, by means of printed "patches". It is not impossible that the revised charts of tomorrow will be produced, on request, by punching a button which will activate a computer to print out a fully up-to-date sheet from digitized data.

Similarly, the conversion, initiated in the late 1970s, of all existing charts (and there are now more than 1,000 navigation charts plus an additional 500 special charts in the CHS catalogue, making this the largest domestic chart coverage in the world), to bilingual, metric formats, may be transformed eventually from a long, arduous, and tedious task to a relatively quick, easy, and automatic process.

The establishment of, and changes in the names applied to geographical features on Canadian charts are handled by the "nomenclature" unit. In the early 1980s the nomenclature unit's prime concern and involvement is the Arctic. There, in one of the few remaining areas of the world where "discovery" still means what it did in the geography and history books of our childhood, names must still be applied to newly discovered features. Hydrographers are human and would, as most humans would, like to achieve immortality in some form or other; not unnaturally, they may hope to achieve this end by naming a new-found feature after themselves.

Understandable as their motives may be, they face some formidable barriers. At the close of each survey, the hydrographer-in-charge submits names

The problem of names

Assigning names to the various geographical features they chart — a task called toponymy — has traditionally been part of the hydrographer's work. While it seems to be a relatively simple task to assign a descriptive name to a feature — Superior Shoal or Western Entrance, for instance — the implications are more complicated. During his years as the regional hydrographer on the Pacific coast, Henri Parizeau worked diligently to untangle the toponymy of west coast charts. The need was obvious: in 1923, to take but one example, a British freighter foundered off the western shore of Vancouver Island in a February gale. The lifeboat crew from Tofino was alerted and at considerable risk proceeded to the scene of the wreck, reported to be Village Island in Clayoquot Sound. The lifesavers found nothing — no wreck, no survivors. And small wonder. The freighter had run aground at Village Island, to be sure, but a *second* Village Island located in Barkley Sound. Fortunately no lives were lost. But the potential for disaster led to renaming the second Village Island as Effingham Island.

for newly discovered features to the nomenclature unit. The names may have been suggested by the hydrographer-in-charge himself, the surveyor who actually found the feature or by the local residents who have called the feature by such-and-such a name for uncounted years.

All names submitted must first be approved by the appropriate provincial member of the Canadian Permanent Committee on Geographical Names. Names of offshore submerged features must gain the approval of a federal committee, currently chaired by S.B. MacPhee, the Dominion Hydrographer. The committee's first criterion is that no name referring to a living person may be applied on a Canadian chart†; tough luck to the hydrographer who discovers a reef and applies to it his own name.

Most of the alterations to existing charts are the result of information gathered from what are called Notices to Mariners. These bulletins, detailing newly discovered hazards to shipping, and changes in aids to navigation such as buoys and lights, are issued weekly in collaboration with, and through the offices of the Canadian Coast Guard. Mariners are required to keep their charts as up-to-date as possible using the most recent edition published and amending them weekly with the information contained in Notices to Mariners. In the words of F.C. Goulding Smith, former superintendent of charts and later Dominion Hydrographer, 1952-57,

> The Hydrographic Service does not approve of the Captain who continued to use his old charts while the new ones were placed carefully under the mattress of his bunk to keep them in good condition. Fortunately, a chart, being made of paper, has a limited life and sooner or later it must be replaced on account of normal wear, tear, and exposure to the elements, however thrifty may be the mariner.

Since 1972 and the ecologically disastrous wreck of the oil tanker *Arrow* off the coast of Nova Scotia, the Canadian government requires all vessels sailing in Canadian waters to carry *approved* charts and publications. In most cases

†Only once since this rule was instituted in 1932 has it been broken. In 1976 Clifford Smith Canyon, located on the floor of the Atlantic southeast of Newfoundland, was named in honour of F.C.G. Smith, Dominion Hydrographer, 1952-7.

the only documents to meet the stringent requirements are those published by CHS. To this end, CHS publications are distributed internationally.

Other publications issued by CHS are *Sailing Directions* and *Small Craft Guides*, and Tide and Current Tables. *Sailing Directions* or "Pilots" are issued in softbound book format about every two years and maintained between editions by Notices to Mariners. They supply information not included on a chart, information such as descriptions (including photographs) of the best approaches to harbours, harbour facilities, anchorages, local history, climatic conditions, rules, regulations, and tables of distances. In other words, *Sailing Directions* are an indispensable companion to the charts. In fact, the publication of *Sailing Directions* in Canada pre-dates that of the nautical chart; Staff Commander Boulton's *Georgian Bay Pilot* of 1885 was published a year before the first Boulton chart of that area. Now *Sailing Directions*, still written by senior hydrographers or experienced mariners and compiled from the information gathered during the field seasons, are available for Newfoundland, Nova Scotia and the Bay of Fundy, Gulf and River Saint Lawrence, the Great Lakes, British Columbia, Great Slave Lake and Mackenzie River, Labrador and Hudson Bay, and Arctic Canada.

Information for the pleasure boater is usually included in *Sailing Directions*, but CHS has in recent years begun publishing a separate series of *Small Craft Guides*. At the moment these are available for the Saint John River, the Trent-Severn Waterway, and there are two covering southern British Columbia. Virtually all are available in both official languages and more are on the way.

Tide and Current Tables are published yearly by CHS in six volumes — for the Atlantic Coast and Newfoundland; the Gulf of Saint Lawrence; the Saint Lawrence and Saguenay Rivers; the Arctic and Hudson Bay; Juan de Fuca and Georgia Strait; and Barkley Sound and Discovery Passage to Dixon Entrance. These tables are necessary aids to the mariner operating in coastal waters where the ability to predict the heights and the times of high and low tides at any given time is essential to safe navigation. Similarly, the information in the current tables in those areas where the times and rates of both slack water and maximum flow of current are predictable (usually when they are affected by the tide) can mean the difference between smooth sailing and disaster.

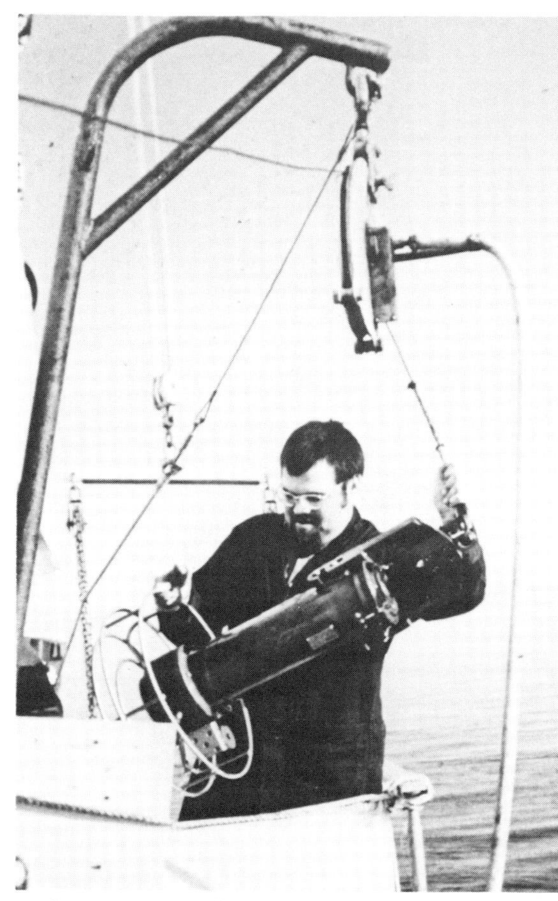

Echo sounders, whether mechanically or electronically operated, bounce a sound wave off the sea bottom and measure the time taken for the echo to return. Knowing the speed of sound in water, and making adjustments for the salinity, temperature and pressure of the water in which he is working, the hydrographer can translate the elapsed time into a measure of depth.

Less concerned with navigational safety and more with the future exploitation of the undersea world are fisheries charts and natural resources maps. These latter are an offshore extension of the national topographic, gravity and magnetic series of maps that cover the land area of Canada. They are produced in cooperation with the Department of Energy, Mines and Resources at scales of either 1:250,000 or 1:1,000,000 to help geologists, geographers, geomorphologists and geophysicists identify those areas off Canada's shores that may harbour potential riches in the form of undersea mineral or hydrocarbon deposits.

Constructed along similar lines and at an even smaller scale — 1:10,000,000 — are those charts produced by the Canadian Hydrographic Service as Canada's contribution to the General Bathymetric Chart of the Oceans, or as it is more familiarly known, GEBCO. The eighteen GEBCO sheets which cover the surface of the world are the fifth edition of an international project begun in 1903 by Prince Albert I of Monaco, and now co-ordinated by the Intergovernmental Oceanographic Commission of UNESCO and the IHO. GEBCO is not meant for navigation but is a research map designed to help scientists, particularly oceanographers, to understand better the history of the Earth, and to expand the boundaries of our, so far, limited knowledge of the two thirds of the Earth which lie beneath the waters.

It is in the development of such projects as GEBCO that the two sciences of hydrography and oceanography meet: the one concerned mainly with the manufacture of an essential product, what R.J. Fraser thought of as "a work of human art", the nautical chart; the other with a search for knowledge for its own sake. Despite their philosophical differences, hydrographers and oceanographers share a common bond in their love of the sea and in their need to spend much of their time away from the land in the pursuit of their different goals. Indeed, in the time before oceanography had attained the stature and the resources it now commands, hydrographers were also oceanographers. Since the beginning of the hundred years of Canadian chartmaking, they have gathered oceanographic information and performed oceanographic experiments while surveying.

Now the two groups share many of the same facilities, including ocean-going ships. In the next chapter we'll look at the ships of the Canadian Hydrographic Service and their evolution to the multipurpose vessels which today provide seagoing laboratories for the study of all aspects of Canada's waters.

Prince of Wales Strait Winter Survey

Winter surveys in the Arctic are carried out from helicopters which leap-frog across the ice in a grid pattern, taking a depth sounding at each stop. The Prince of Wales survey followed this pattern and was a survey of part of the Northwest Passage route. Far from civilization, the survey was a test of men and equipment against the sub-zero temperatures during the short hours of daylight of the Arctic winter. At an isolated Arctic base camp Arctic fox, musk ox and caribou are frequent visitors.

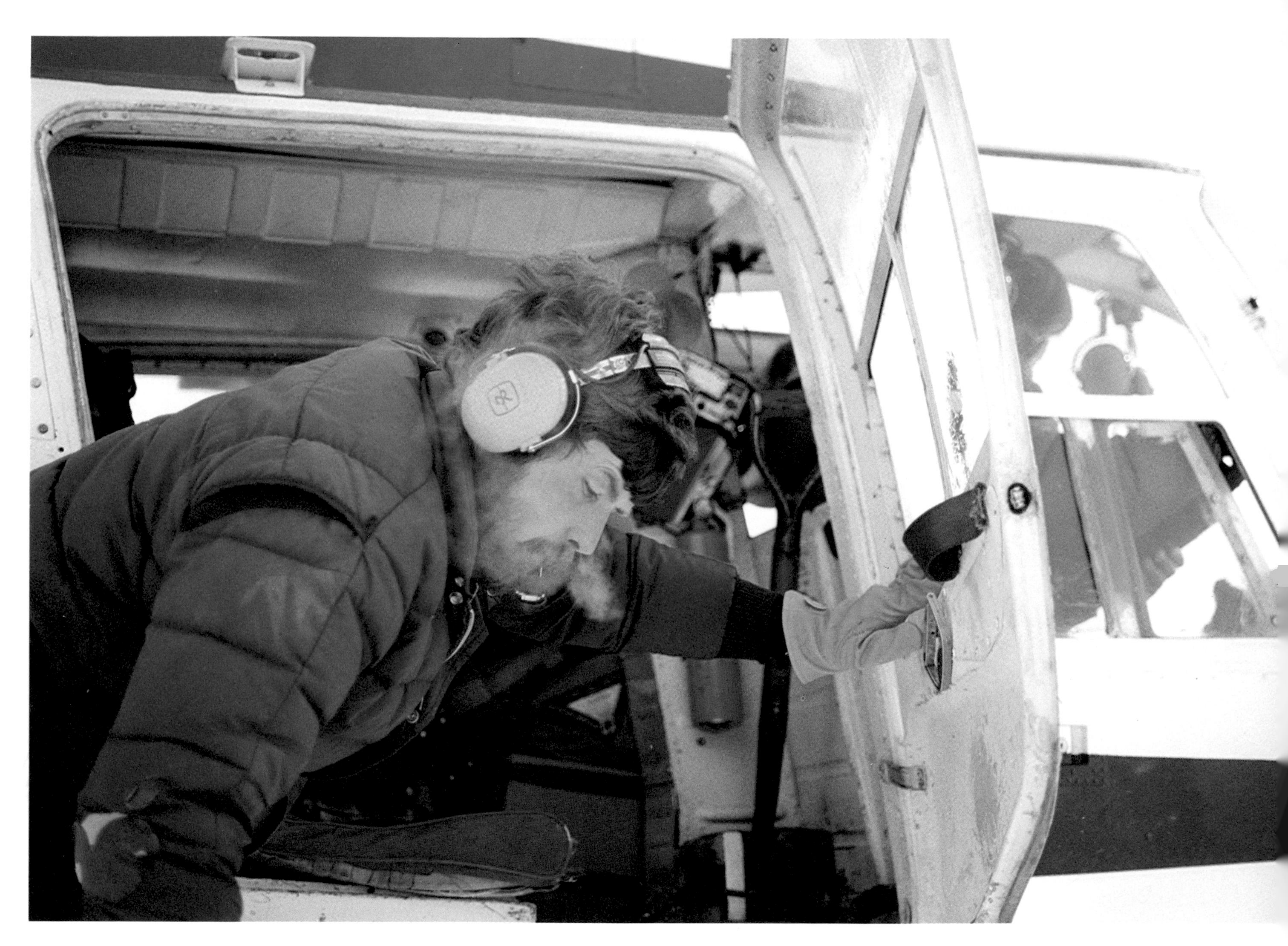

QUASAR
C-FXSK

The Ships

We sailors are jealous of our vessels. Abuse us if you will, but have a care for what you may say of our ships. We alone are entitled to call them bitches, wet brutes, stubborn craft, but we will stand for no such liberties from the beach.

Sir David William Bone
Merchantmen-at-Arms, 1919

THE MEN OF THE CANADIAN HYDROGRAPHIC SERVICE HAVE seldom been so sensitive or sentimental about their ships as was Sir David Bone. For the most part, they've lacked both the opportunity and the motivation. Hydrographic ships since the time of Boulton and the Georgian Bay Survey have more often than not been vessels of convenience — chartered, borrowed, bought second hand, rarely built especially for the use of Canadian hydrographers. The men themselves, since the days of William J. Stewart in the late nineteenth century right up until the present have been mainly graduate engineers or surveyors to whom sea legs were grafted over the course of several survey seasons.†

But the innate romance of the sea has caused exceptions to these rules to be better remembered. The major ships such as the *Acadia*, the *Wm. J. Stewart*, the *Baffin*, all of them created for the service of hydrographers have, together with their sailing masters and crews, become the stuff of legend in the CHS.

When Boulton began the Georgian Bay Survey in 1883 it was with vessels begged and borrowed from the local residents. The first ship actually purchased for the surveying service arrived on Georgian Bay in the summer of 1884. She

†Often enough, the "grafting" took place in the opposite direction: experienced mariners learned the surveyor's arts. At the end of the World War II many officers who had served with the Royal Navy, RCN, and the merchant marine of both Britain and Canada joined the service. They were men who had used charts, knew them and appreciated their importance. Many of the recruits left the service within a few years but enough remained to leaven the engineer/surveyor mix.

The well kept secret of Superior Shoal

Lake Superior enjoys a reputation as the deepest of the Great Lakes, a renown it deserves. Early charts of the lake showed its middle depths to be in excess of one thousand feet. Most mariners, trusting the charts' accuracy, saw little need for extra precaution while sailing well offshore. Over the years, though, more and more commercial vessels disappeared under mysterious circumstances; during World War I, for instance, two wooden mine sweepers built in Port Arthur (Thunder Bay) for the French government vanished without a trace on their way downstream to Sault Sainte Marie. Gradually tales and rumours — and eventually confirmation — of the so-called Superior Shoal began to circulate in shipping circles. In fact, the presence of the shoal, later described by former Dominion Hydrographer R.J. Fraser as "virtually a submerged mountain, rising out of the blue depths to about 21 feet of the surface," had been an open secret for years. The crews of American fishing tugs out of such Michigan ports as Grand Marais were well acquainted with the shoal since in the shallow depths above the shoal swam large schools of lake trout and siscoes, and the fishermen were in no hurry to broadcast the news to competitors.

La Canadienne's ignominy

On 20 June 1912 the survey ship, *La Canadienne*, entered Lock 22 of the Welland Canal upbound for a season's work on Lake Superior. She was still underway while the lock gates closed behind her; fore and aft, seamen on the lock walls were snubbing her down with hawsers to hold her steady as the lock filled. One seaman, a London Cockney presumably with scant experience, insufficiently secured his line. The ship continued to move ahead and crashed into the lock gate ahead. The weight of water in the higher lock sprung the gate, and the full mass of the water poured down on *La Canadienne*, driving her into the wall and opening a hole in her bottom. She sank immediately. Two young men, watching the ship proceed through the canal from a vantage point on the lock wall, were swept away by the wave and drowned. R.J. Fraser, who was appointed Dominion Hydrographer in 1948, was with the ship's hydrographic staff. Later he was to recall that "we were reimbursed for the loss [of] our cabin contents, though some of the items of clothing. . . such as a dress suit and a quantity of silk socks, etc., were questioned by the department and disallowed. They were ruled to be clothing not required and unsuitable for work on the north shore of Lake Superior."

was a twenty-year-old American tug. Originally christened *Edsall*, she was immediately renamed *Bayfield* in honour of Boulton's predecessor on Georgian Bay, the pioneer nautical surveyor, Admiral Henry Wolsey Bayfield.

Even after she was condemned as unfit for duty, the 100-foot *Bayfield* continued to work the waters of the Great Lakes until her retirement in 1902. Despite her age, her generally dilapidated condition, and her decidedly unglamourous appearance, the *Bayfield* did her job well for almost twenty years. Her sturdiness and her forgiving nature in the face of the hazards of her occupation — the bad weather, hidden rocks, shoals and regular steaming into unknown waters — were qualities that set the standard for the workaday ships of the CHS that would follow in her path. The phrase "sturdy little ship" is used even today when hydrographers are pressed for some laudatory comment about their vessels.

Bayfield's successor on the lakes was another ex-tug, albeit a larger — at 140 feet — more powerful, ocean-going one. The *Lord Stanley* was twelve years old when purchased for the Great Lakes Survey in 1901. Meant to replace the *Bayfield* the next year, she was damaged in Toronto Harbour after refitting and didn't see service on the lakes until 1903, at which point the *Bayfield* was retired and that name passed on to the *Lord Stanley*. The rechristened *Bayfield* (II) maintained the tradition of longevity and did not retire until 1935. Her most famous survey was one of her last, the sounding of Superior Shoal in the main shipping channel of Lake Superior in 1930.

By 1905 the Canadian government had assumed full responsibility for the charting not only of its inland waters but for all the nation's lengthy east and west coasts. Unfortunately, with no survey ships in these waters the proposition remained academic at best — laughable at worst with possibly tragic consequences if the situation were not corrected. Accordingly, William J. Stewart, newly named chief of the Canadian Hydrographic Survey, requested in the fall of 1905 that the Department of Marine and Fisheries undertake the construction of two new survey ships — one for each coast.

The following year a twenty-six-year-old, 154-foot steamer, *La Canadienne*, was purchased from the Fisheries Protection Service and refitted for hydrographic work. She replaced the tidal survey steamer *Gulnare* (a former

The SS Lord Stanley *was a 140-foot ocean going tug when she was purchased by the government in 1901 to replace the aging CGS* Bayfield. *Refitted at Toronto as a survey vessel she entered Canadian hydrographic service in 1903 as CGS* Bayfield *(II) and remained in active service until 1930.*

survey ship chartered by the Royal Navy) which had been performing some hydrographic duties on the lower Saint Lawrence and east coast.

On the west coast in 1908 Stewart's wish for a brand new ship was realized. The *Lillooet*, built in Esquimalt at a cost of $150,000, was the first vessel designed and constructed especially for the Canadian hydrographic fleet. At 172-feet, with a gross tonnage of 575, she could accommodate six hydrographers, carry four or five smaller boats and a total crew of about forty.

Birds and other diversions

During the 1908 survey season on Lake of Two Mountains in the Ottawa River, the hydrographers used a steam launch for transportation and the houseboat *L'Arche* as living accommodation. The boat provided considerably more space and comfort than the normal cramped quarters aboard ship, and was a great improvement on life under canvas, the usual means of housing shore parties. In fact, *L'Arche* was said to provide "all the comforts of home". Her accoutrements included a pet canary, the pride of one of the engineers. Attending to the bird and to the "chief manufactured product of Berthierville" were said to be the engineer's chief preoccupations. Berthierville, Quebec, was the nearby home of a large distillery.

The first ship specifically constructed for Canadian hydrographic work was CGS Lillooet, *shown here at her launching in Esquimalt in 1908. At 172 feet, she was the work horse of the Pacific region for more than twenty years.*

Houseboats like the Pender, *shown here, served as living quarters and plotting room for the hydrographers, who put off each morning for their survey work ashore or in launches.*

For the next twenty-four years she was virtually the only hydrographic ship on the west coast, performing without incident until replaced by the *Wm. J. Stewart* in 1932. The *Lillooet* was able to extend her sphere of operations, however, by the use of houseboats towed behind the ship. These barges, among them the *Pender*†, *Fraser*, and *Somass*, could carry a couple of launches, two or three hydrographers, and crews of fourteen to seventeen men.

Again on the east coast, in 1910 the hapless *La Canadienne* was replaced by the stalwart *Cartier*, a 164-foot steamer designed by R.L. Newman of Montreal but built in England. In appearance and character, almost a sister ship to the *Lillooet*, *Cartier* surveyed east coast waters until 1945 (with wartime secondments to the navy for patrol duty). But this fine ship was overshadowed in 1913 with the arrival of what became known as ''Canada's grand old ship'', the *Acadia*.

Constructed by the same firm as built the *Cartier*, the 170 foot, $330,000 *Acadia* was a beautiful, graceful ship, the pride of the Canadian Hydrographic Service for much of her fifty six year career. Retired in 1969, she now is docked at the Maritime Museum of the Atlantic in Halifax and is open to the public. One quick tour through the ship and it's immediately obvious why *Acadia* was, even as she advanced into old age, regarded with respect and admiration by all who sailed in her.

Built in the Edwardian tradition of her unfortunate contemporary, the *Titanic*, *Acadia* exudes an air of quiet dignity and conveys a sense of permanence in every part — from the steel plated hull, to the gigantic fittings of her coal burning, steam engineroom, to the splendour of the senior hydrographer's cabin where one is surrounded by polished teak, mahogany, brass, and the luxury of horsehide banquettes and easy chairs. The hydrographers' mess resembles

†*Pender* lives on — or, rather, has been reincarnated — in both name and function. Since the 1980 survey season a relatively new barge bearing her predecessor's name is again in service on the west coast as a floating home and office for hydrographers. The new *Pender*, some ten by twenty metres in dimension, was built as a tender for the submersible *Pisces*; it carries a covered hangar into which the launches — she carries three — can be hauled out for repair, and is staffed by a crew of about twelve including six hydrographers and the coxswains who man the launches. A cook and steward are also carried.

The Chrissie C. Thomey

The *Chrissie C. Thomey* was a veteran merchantman engaged in the Newfoundland-West Indies sugar and rum trade when she was purchased for hydrographic work in the Arctic. Wooden hulled and 200-feet-long, she carried no auxiliary power and was driven by sail alone. She carried three gaff rigged masts, and fully dressed with jib and tops'ls flying, she looked more like a racing schooner than a survey ship. The lack of an engine was her undoing; caught behind a sandbar at Rupert River, James Bay, she might have escaped had she been able to power out at high tide. Tacking under sail she was doomed and was abandoned during the spring break up in June 1913. She was the first ship casualty of the Canadian Hydrographic Service.

The *Chrissie C. Thomey* is shown, right, on the ways during her refit as a survey vessel. Captain Thomas Gushie, a former sealing master from Newfoundland was her captain.

Under schooner rig, the Chrissie C. Thomey *(far right) could make fifteen knots.*

There were no shipyard facilities in Hudson Bay but the crew (right) repaired the boom under the watchful eye of Black Tom.

To his crew, Gushie (above) was known as Black Tom O'Brigus.

. . . the hydrographers' quarters [aboard CSS *Wm. J. Stewart*] were spacious and, in their style, comfortable, though as electrical equipment was added throughout the years and placed a greater load on her generators the wattage of the cabin lights was reduced and an underground war developed between the surveyors who slipped in brighter bulbs and the engine room staff who just as regularly replaced them with 25 or 40 watt bulbs.

R.W. Sandilands
in *Lighthouse*, Journal of the
Canadian Hydrographers Association
November 1979

the dining room of a fine hotel complete with china, silver, linen, crystal embossed with the ship's crest, and call buttons at each chair with which a uniformed steward could be summoned.†

With the two ships, *Cartier* and *Acadia*, the CHS established itself as a proud, visible, and important presence on the east coast. *Acadia* spent her first two seasons charting the Hudson Bay route in Canada's north. Her thick steel hull was specially designed to withstand the rigours of Arctic ice. Also she was the first CHS vessel to be equipped with wireless radio and, in an effort to compensate for the extra sensitivity of compasses operating in proximity to the magnetic North Pole, all metal within fifteen feet of *Acadia*'s compass was non-magnetic sceptre bronze. She returned to Hudson Bay several times but spent most of her years surveying the waters of the Gulf of Saint Lawrence, the coast of Nova Scotia, and, after 1949, the rugged coastlines of Newfoundland and Labrador.

Acadia was one of the first government ships involved in the regular gathering of oceanographic information, starting in 1914. When it came time, in 1931, to design a ship comparable to the *Acadia* in stature and status, for the use of west coast hydrographers, some oceanographic facilities were included in the plans.

That ship, the *Wm. J. Stewart*, was built in Collingwood, Ontario, sailed east then south and through the Panama Canal, arriving at Victoria in July 1932. She was the largest ship constructed to that time for the CHS — 228 feet long, and 1,295 gross tons. Her total ship's complement was sixty-eight, including eight hydrographers, eight ship's officers, and a crew of fifty-two (which included, during World War II, seven female cooks, stewardesses, and laundresses). By the summer of 1933 the *Stewart* had been outfitted with the latest in hydrographic equipment and she took over from the retiring *Lillooet* the enormous task of charting Canada's Pacific coast.

†*Acadia*'s most noticeable refinements were handsome carved bow badges of teak with ornate gilt curlicues. During a winter's layup in the 1950s at her operations base of Pictou, Nova Scotia, the carvings were removed and placed in storage to permit work on the ship. Fire destroyed the storage shed and the carvings, which were never replaced.

The "Willie J." was the pride and joy of CHS's Pacific region from her arrival in 1932 until her departure in 1975. Officially the CGS (later CSS) Wm. J. Stewart, *she was 228 feet in length and carried eight hydrographers and a crew of sixty. Named for William J. Stewart, the ship has been berthed since her retirement at Ucluelet on Vancouver Island's west coast where she serves as a floating hotel.*

(Like all federal government ships, the *Stewart* was registered in Ottawa and carried her port of registry on the stern beneath her name. Many a gullible visitor to the west coast, seeing the ship and her home port notation, was regaled with tales of the incredible problems encountered in sailing the ship west from Ottawa, through an intricate system of interconnecting waterways across Ontario and the western provinces, to her operating base in Victoria.)

During World War II, the *Wm. J. Stewart* was the only hydrographic ship engaged in surveying on either coast, all others having been commandeered by the military. However, on 11 June 1944 this work was interrupted when *Stewart* hit the notorious Ripple Rock in Seymour Narrows. Salvaged from the muddy waters of Plumper Bay, named for an earlier survey ship, HMS *Plumper*, *Wm. J. Stewart* was towed back to Victoria, repaired, refitted and was back

The Acadia

When the Canadian Government Ship (latterly renamed the Canadian Survey Ship) *Acadia* entered into the service of Canadian hydrographers in 1913 she represented state-of-the-art technology. She was the first Canadian survey ship to carry radio communications, the first to use the gyro compass, the first to regularly use gasoline launches (although previous ships had tested them), the first. . . the list goes on.

She was thoroughly competent to her task but she was more besides: a subsequent generation might have called her "user friendly." She was more than merely comfortable; her accommodations were *elegantly* comfortable. It was often said of *Acadia* that she more closely resembled a first class hotel than she did a survey ship. During the glory years her elegance was epitomized by the intricately carved teakwood plaques on her bows.

On *Acadia*'s maiden hydrographic voyage she sailed in 1913 from Halifax to Hudson Bay. Several hydrographers were on board, among them one young man named Lloyd Prittie: Prittie assumed responsibility for recording the survey photographically.

He was a talented amateur and during the voyage exposed some one hundred or more negatives. On Prittie's death his widow donated his photographs to the Maritime Museum of the Atlantic, Halifax, where *Acadia* —although retired —lives on as a reminder of days long gone in the Canadian Hydrographic Service. These pictures are from the Lloyd C. Prittie Collection.

The muted elegance of the hydrographers' wardroom aboard Acadia *was more akin to an exclusive men's club than to the practical but spartan facilities on most survey ships. Wood panelling lined the walls, decorative etched glass was set into the doors, white linen graced the table. Flatware and china carried the ship's crest, and a call button at each place summoned the steward. Needless to say, the hydrographers dressed for dinner — not black tie, perhaps, but ties nonetheless, and jackets. Lloyd Prittie, the hydrographer who made these photographs, sits third from the left. His left hand is concealed behind his neighbour's back, possibly to keep out of sight the cable release with which he tripped the camera's shutter.*

Just before her departure for Hudson Bay in 1913, Acadia *(far right) lies at dockside, taking on supplies, probably in Halifax Harbour with an industrial smokestack in the background. Moored ahead of* Acadia *is a sailing steamship, powered with an engine but rigged with masts and yards as well so that sails can be hoisted in an emergency.*

The photographer identified the hydrographer in this picture (right) as "Chambers." He stands on the foredeck of one of Acadia's *gasoline launches, boathook in hand, about to bring the launch to the ship's side. The ships seen in the right background, and the lack of a cabin on the launch suggest this photograph was made before* Acadia *sailed north.*

At an Arctic settlement this Eskimo kayak attracted Prittie's photographic attention.

Once in the Arctic, Acadia's *launches were fitted with protective cabins to shelter hydrographers and their equipment. The helmsmen of the canoe-ended craft, however, remained exposed in the open cockpit aft.*

Hemmed in by Arctic ice off Cape Chidley on Labrador's northernmost tip, Acadia *waits for the floes to open. The delay gives her hydrographers time to go "ashore."*

Our rocky coasts

Although it's something of an embarrassment to admit it, hydrographers are sometimes those most in need of the charts they produce. On 11 June 1944 while surveying waters on the Pacific coast, CSS *Wm. J. Stewart* struck the notorious Ripple Rock in Seymour Narrows near the southern end of British Columbia's inner passage to the north. Ripple Rock had claimed many another victim in the past, but that was small consolation to the hydrographic community. There were no fatalities among the fifty-eight men and seven women in the crew, and the ship was successfully beached with the entire complement's help. Later, she was refloated and towed to Victoria for repairs. "The hull stood up well to the accident but the engine room was badly damaged, the electrical equipment was ruined and the beautiful maple panelling of the cabins and public rooms had to be completely removed.": Sandilands. *Stewart* was back in service the next year and remained active in surveying the west coast until she was decommissioned in 1975. In 1958 in the largest man made, non-nuclear blast to the time, Ripple Rock was dynamited out of existence.

in action by the summer of 1945. She remained in active service until 1975.

The years from 1932 until the beginning of World War II were the glory days of ships in the CHS. By war's end all of the vessels, in particular the showcase ships *Acadia* and *Stewart*, were exhibiting evidence of their age and/or misadventures. But at the end of the war, Canada found herself overstocked with warships not required for a peacetime navy. The post-war years brought an expansion in CHS's activities and with it the need to expand the fleet; both purposes were accomplished with the secondment of surplus warships.

Parry (ex *Talapus*) had been a coastal patrol vessel, 87 feet long with a wooden hull; she was converted to a hydrographic ship and used mainly in current and tidal surveys. *Marabell* had been a USN minesweeper (YMS – 91) purchased in 1948 by Doctor Ballard of pet food fame and converted to a motor yacht; in 1953 she was sold to the CHS and refitted for hydrographic work. *Ehkoli* was another patrol boat similar to *Parry* and was refitted for oceanographic surveys. All three ships remained in service on the west coast until 1969 when they were replaced with newly constructed vessels.

The *Cartier* was another wooden hulled ship built as a coastal minesweeper for the Russian Navy in shipyards at Collingwood, Ontario. However, before she slipped down the ways in 1945 the war in Europe was over and she was never delivered. In 1947 the CHS bought her and from 1948 until 1962 she surveyed the southern Labrador coast and in the Gulf of Saint Lawrence. From 1962-4 she served on the Great Lakes as a hydrographic training ship after which she returned to active survey work on the Lakes until 1968 when she was retired.

At the same time on the east coast two RCN Algerine class minesweepers, *Fort Frances* and *Kapuskasing* were demobilized and refitted for hydrographic surveys.

But even with the addition of these former warships to its fleet, CHS could not keep up with the post-war demands for new charts. Because demand for ships was high, and because the requirements of wartime had diverted shipyard production from commercial to military vessels, suitable vessels were in short supply.

The impasse was broken by CHS, when it could, chartering Newfoundland sealing ships; the sealing season is finished in late winter so the ships were available at exactly the time of year when CHS required them. Then too, the sealers were constructed for ice work — sturdy timberwork forward with bows sloped sharply aft so they tended to ride up over broken ice and force it out of the ship's path. Further, the sealers' officers and crews were experienced in Arctic and sub-Arctic waters where many of the new surveys were required.

At the end of the sealing season the ships chartered for that year's work were sailed to the nearest shipyard in Newfoundland or on the Nova Scotia coast and refitted for the hydrographic season. This entailed installation of a drafting room, conversion of sealers' quarters to the hydrographers' use, and

The Parry *(top),* Marabell *(middle) and* Ehkoli *(bottom) joined the Pacific region fleet after World War II. All had seen wartime service,* Parry *and* Ehkoli *as patrol ships,* Marabell *as a minesweeper.*

As on the Pacific coast, the Atlantic region acquired former warships and converted them to hydrographic work. Fort Frances *(shown here) and her sister ship,* Kapuskasing, *had seen service with the Royal Canadian Navy as minesweepers.*

a thorough scrubbing-down of topsides and interior though the stench of seal flesh still pervaded the whole ship and often enough was still noticeable at the end of survey season.

North Star IV was one such chartered sealer. (Others were *Terra Nova*, *Theta*, *Theron* and *Algerine*.) In August 1960 she was on survey in James Bay when a storm blew up. Mike Bolton, later the regional hydrographer on the Pacific coast, was the hydrographer-in-charge of the survey. He relates what followed:

We were sailing in uncharted waters so I had stationed a launch ahead of

> the ship to take soundings. But the wind continued to increase — the waves became huge — and I had to recall the launch. The ship continued to move ahead, slowly, when suddenly there was a hell of a crunch . . . we were hung up on an uncharted rock.

Reg Lewis was a young hydrographer aboard on his first survey. When *North Star* grounded, "I thought it was the end of the world. I've never forgotten — nor shall I — my first hydrographic survey."

> We would probably have saved the *North Star* [Bolton continues] if the master hadn't panicked. The tide was rising; when the wind died down a little I ordered the launches lowered to lighten the ship; with a little patience and a little luck I believe *North Star* would have floated free. But patience . . . the master didn't have it. He signalled the engineroom for full speed ahead, then full speed astern. He repeated this often enough that he literally sawed a hole in the hull.
>
> We got the whole ship's complement off in the launches — no loss of life, no injuries — and we camped ashore overnight. By next morning the sea had battered the ship to pieces. I suppose the local Inuit might have salvaged some of her timbers.

North Star Shoal off James Bay's eastern shore commemorates the accident.

In the post-war years CHS arranged with the federal Department of Transport (DOT) to place sounding launches aboard DOT icebreakers serving the Arctic. As well the icebreakers were equipped with a hydrographic plotting office. CCGS *C.D. Howe* was the first used (in 1952) in this fashion and each year thereafter as many as three teams of hydrographers were assigned — and still are, in fact — to serve aboard the Canadian Coast Guard ships.

Not unnaturally the Coast Guard's prime responsibility lies in non-hydrographic directions so the success of the hydrographic endeavours carried out on its ships varies from season to season depending on the ice conditions in the Arctic and other demands placed on the ships.

Interestingly enough one of the most startling initial contacts with what may be the western Arctic's most curious sea bottom anomaly was made by a hydrographer serving aboard CCGS *Sir John A. Macdonald*.

Lessons learned the hard way

Hydrographic ships — at least in Canada — are limited by the weather to an effective survey season of about five to seven months. The rest of the year their usefulness as survey ships is "wasted". That, at least, used to be the view of many in the various departments of the federal government under which the CHS has operated. And the yearly layups rankled — "inefficiency" was the word. In 1922, over the objections of the then Chief Hydrographer William J. Stewart, the Department of Marine and Fisheries decided that CSS *Acadia* should be used during the winter months as an icebreaker in harbours along the southwestern shore of Nova Scotia. *Acadia* had been built to ride up over floe ice and crush it with her own weight, but she was not constructed to the task of breaking up sheet ice. On 16 March 1923 while engaged in this task *Acadia* damaged her rudder post severely. Further icebreaking was postponed and the experiment was never repeated.

CSS Baffin *in Trinity Bay, Newfoundland*

The major survey vessel for the CHS on the East Coast, CSS *Baffin*, works on a number of vital surveys along the coast of Newfoundland. Here hydrographers and crew are shown working in Trinity Bay.

4

1
AA7334

When the captain lost his head

While surveying the north shore of Lake Superior on 23 June 1908 the *Bayfield* ran aground on McGarvey Shoal near the town of Schreiber. The skipper, who was on the bridge at the time, was apparently concentrating more on picking up the sounding gigs than watching where he steered the ship. *Bayfield* with some of her crew aboard was towed to Collingwood on Georgian Bay for repairs. One night during the enforced hiatus several of the hydrographic staff spent the night in town, probably, as seafaring men are wont to do, drinking. On their return to the ship one of the men "liberated" a headless tailor's mannequin used as a display outside the shop. Smuggling it aboard *Bayfield* he hoisted it to the masthead on the flag halyard. The following day the captain discovered the prank, and interpreted it as a personal criticism of himself for having lost his head in the crisis. He fumed, he stormed, he threatened dire punishment for the culprit. The jokester was never discovered. The captain lost his command the following year.

In 1969 the presence of petroleum deposits beneath the Beaufort Sea had been confirmed. Canada and the United States, with an eye on the future exploitation of the oil and natural gas, determined on a test to see if a huge tanker, sufficient in size to make removal of the hydrocarbons practical, could negotiate the Arctic, entering from the Atlantic and sailing through the Northwest Passage to Prudhoe Bay in Alaska. The task was assigned to MV *Manhattan* to be accompanied by the icebreaker *Sir John A. Macdonald*.

Through most of the voyage the *Macdonald* went ahead, breaking the pack ice in the *Manhattan*'s path and using her echo sounder continuously to check bottom depths. As the two ships cleared Prince of Wales Strait and headed across the Beaufort Sea, *Manhattan* moved for the first time into the lead.

Ken Williams was hydrographer-in-charge aboard *Macdonald* (later he was named regional hydrographer at Quebec City) and he was busy in the plotting room when he noticed with horror that his echo sounder indicated that the bottom was shoaling drastically. He grabbed his telephone, dialled the bridge and spoke — practically screamed — at Admiral Tony Storrs, Coast Guard Commissioner, "Get down here . . . quickly!"

When Storrs appeared Williams gestured at the echo sounder and said, "Look at that — just look. Do you see what I see?" Storrs stabbed his finger at the sounding graph pointing to the alarming peak traced out by the stylus. "You mean *that*?" he asked, poking his finger through the graph paper. Storrs also managed to damage the stylus and put the sounder out of commission entirely.

So it was that the first submarine pingo† was discovered in the western Arctic and named "Admiral's Finger". Williams replaced the stylus and repaired the

†Previous to this discovery, pingos were a well known land feature dotting the barrenlands of the Northwest Territories. They are abrupt hillocks rising as high as thirty-five or forty metres from the surrounding flats and often measuring three hundred metres in diameter at their base. They have been described as looking like "giant anthills" and have a core of ice covered with dirt, shale and rocks. Submarine pingos, like the Admiral's Finger and the almost one thousand more that have been discovered since 1969, seem to be similar in size and in possession of an ice core. Of the hundreds found and plotted on charts only one percent represent a hazard to shipping. These, with a minimum water depth of twenty metres or less above them, protrude into the danger zone of a supertanker's hull.

machine but today says that if Admiral Storrs "had been one rank lower I'd have bopped him."†

The emphasis that had been placed on Arctic exploration in the immediate post-war years continued and continues today with even greater urgency.

In addition to chartering ships and placing men and equipment aboard DOT icebreakers, the CHS, in 1956, took delivery of its largest hydrographic ship. CSS *Baffin*, a 285-foot vessel of 3,700 ton displacement, was constructed specifically to meet the requirements of Arctic charting. But the age of hydrographic luxury afloat — as represented by *Acadia* — was gone forever and *Baffin*, while comfortable, is not the Ritz. The hydrographer-in-charge still occupies his own suite — a largish stateroom/office, a bedroom and his own bathroom — but gone is the wood panelling and custom fittings of the *Acadia* era.

As the *Acadia* was in her day a testing ground for new technologies such as the gyroscope compass and the echo sounder, so the *Baffin* was one of the first hydrographic vessels to utilize the new electronic positioning systems. Her hydrographic plotting room was then one of the most advanced of its kind and in the years since has been outfitted with an array of computer terminals and allied technology.

Unfortunately, *Baffin*'s first test of the new electronic navigational aids — it happened to be Decca — went awry, an incident described in greater detail in the next chapter. *Baffin* and Decca survived the "bad day at Black Rock" but not without some intensive soul-searching and the creation of some serious doubts about the efficiency of the new electronics.

In the years since the abortive first trial the electronics have proven themselves and allayed the doubts. *Baffin* has shown herself to be a survey ship *par excellence*. But it's generally acknowledged that she is probably the last

†The tale is not quite ended. When he'd repaired his sounder Williams contacted *Manhattan* on the radio telephone. "I asked them if they'd passed near the shoal that had shown up on our sounder," Williams recalls. "The reply was : 'Hold on a moment, I'll check.' Obviously they didn't have anyone on sounder watch." In a few minutes a fuller answer was received: "Holy Christ!" it began prayerfully or profanely. "Yes we did." And it was accompanied by a request that *Macdonald* resume the lead; she did and did not relinquish it until the ships had departed the Arctic.

Those in peril on the sea

The sea is an unforgiving mistress and hydrographers — like all who set sail — must be constantly on guard. It is a mark of the high standards of safety enforced in the CHS that so few fatal accidents have happened in the service's history. One death occurred in September 1960 when a dory capsized during a squall while attempting to land on Kunghit Island on the west coast of the Queen Charlotte Islands. Hydrographer Ralph Wills, coxswain Olav Watne and seaman Robert Lee were pitched into the sea. Wills managed to grasp a floating log and stay afloat until he was rescued by a survey launch. Lee grabbed hold of an offshore rock but was later washed from it and cast ashore where he spent the night in company with two other men from another dory that had capsized in an attempt to reach him. Watne disappeared and his body never recovered; it was assumed he had been knocked unconscious when the dory overturned. The young Norwegian immigrant was eulogized by Captain Howie Matheson as "a true Viking, a natural born seaman".

The Chrissie Thomey *is abandoned*

In the summer of 1912 the hydrographic ship, *Chrissie Thomey*, departed Halifax for the 2,800-mile voyage to James Bay. She was a sleek 100-foot, wooden merchantman. Deep keeled, she looked more like a yacht than a scientific ship, and her looks were not deceptive — in a stiff breeze her schooner rig could churn out fifteen knots. She held the course record for the Saint John's-West Indies run. But she carried no auxiliary power. She was brought to winter berth at the mouth of the Rupert River in James Bay, clearing a sandbar at the entrance by inches. Indeed, had it not been for the highest tide of the year, combined with a stiff wind that piled up the water even higher, she never would have made it. She was never able to leave. The combination of tide and wind that had permitted her entry never recurred; with an auxiliary engine — or under tow from a powered boat — she might have negotiated the narrow channel. Tacking under sail she could never have made it.

of a breed. With the emergence since the mid-1960s of increasing oceanographic requirements for ships with many different types of facilities and equipment, added to the restraints of federal budgetary policies, the large dedicated hydrographic ship has become a relic of a different, less complex past. Newer vessels such as the *Parizeau* and the *Vector* on the west coast, and the *Hudson* and *Dawson* operating in the east and in the Arctic, are multipurpose ships, equally able to carry a hydrographic survey party or an oceanographic team.

In the early 1960s two new small ships joined the hydrographic fleet. CSS *Richardson*, a 60-foot steel hulled vessel designed to work and winter in the Arctic, was assigned to the Victoria, British Columbia, office (now relocated at Patricia Bay). For several years her combined master/hydrographer-in-charge was Captain T.D.W. McCulloch. In spite of her strengthened hull *Richardson* barely escaped being crushed in ice in 1967 off Point Barrow, Alaska.

The second ship was CSS *Maxwell* which carried two 26-foot sounding launches; she became the "fire engine" of the Atlantic Region, dashing into action when the urgent need for a survey arose.

The future of the purely hydrographic vessel appears to lie in the direction of the modern launch — the descendant of a long line of relatively small, fast and, yes, "sturdy" boats that have been used by hydrographers since the early years of the century.

The hydrographic launch of today, though it is a far cry from the sailing gigs used by James Cook when he began to organize the arts of chartmaking into the science of hydrography, is the direct descendant of those early boats.

Cook's gigs were merely the ship's boats used for a multitude of purposes — for rowing the crew ashore when the vessel was moored offshore, to take one instance. They carried "steps" or brackets into which a mast could be inserted and a sail raised, but the primary motive power came from oars, usually four or six, single-banked, two or three to each side. The gigs were constructed, of course, of wood.

Used for surveying, the gigs were rowed close to shore or the shoal area to be charted. They were then rowed along the sounding lines while the hydrographer, standing in the stern sheets with his sounding board, sextant, notebook, fieldglasses, and logbook, steered the boat, took position fixes of the sound-

CSS Baffin *(top) and* CSS Richardson *(lower left) were added to the CHS fleet in the 1950s and '60s.* Baffin, *commissioned in 1956, was – until her refit in 1982 – the supremely complete hydrographic ship operating in Canada; she now functions as a combined hydrographic-oceanographic vessel.* Richardson *is based on the west coast and was specifically designed to winter in the ice of the western Arctic.*

Any system of surveying a river forces the surveyor to sound in parallel lines, crossing and recrossing. The steamer [*La Canadienne*] can, at best, in smooth weather, steam eight knots. The tide frequently runs four knots, so it may be seen that she is unable to keep on course directly at right angles to the . . . river . . . in changing from line to line upon the completion of one, the steamer is hardly able to make headway.

William J. Stewart
Chief Hydrographer
in his annual report, 1907

ings, watched the leadsman's line, and recorded the soundings. Obviously the invention of the gasoline launch meant not only a saving in manpower but also, in the long run, an increase in efficiency on the part of the harried hydrographer.

Gasoline launches were first used by the CHS merely to tow the sounding gigs. But by 1910 the hydrographic schooner, *Chrissie Thomey*, was equipped with two gasoline launches. One of these, the thirty-foot *Nelson* was affectionately known as the ''Sea Louse'' because, in the words of former Dominion Hydrographer R.J. Fraser, ''one never knew where it was going to turn up for a job of charting work. . . Not an elegant or refined name, but descriptive.'' About the same time, on the west coast, a launch named *Budge* was nicknamed (for obvious reasons) the ''Never Budge'' by her crew.

Perhaps because of their reputation for being undependable, the use of gasoline launches was not officially approved by CHS headquarters in Ottawa until 1913. Besides thinking them unreliable, William J. Stewart believed the launches to be extravagant because they were more expensive to lease and buy than the old gigs, they required the purchase of petrol, and also the use of a gasoline engine demanded the presence of a qualified mechanic in the event of anticipated breakdowns. That year Henri Parizeau, having seen most of the crew of *La Canadienne* jump ship for better paying jobs in Fort William, and realizing that there remained not enough men to handle gig sounding, wired Stewart in Ottawa for authority to rent gas launches. Permission was granted, ''reluctantly'', as Parizeau noted sourly.

After that, gasoline launches, most of them in the thirty-foot category, gradually replaced the sounding gigs as standard equipment on the larger hydrographic ships. And during World War II most of the survey work was accomplished using only launches as the ships were seconded to the navy.

In 1941, hydrographer H.L. Leadman, working from the launch *Henry Hudson* wrote that:

> During the summer I noticed that there were a great many sunny days when our 27 foot launches were driven to shelter and the day was a total loss as far as sounding was considered. They could not stand up to the fresh summer breeze. In replacing these launches as they wear out it would be highly advisable to use a much larger craft. I would suggest one about 40 to

> 45 feet long, under four feet draft, one of easy running and fairly light construction. In view of our Cape Breton experience and others, it would have two engines of about 40 HP each.

Interestingly enough, a launch very similar to what Leadman described appeared the very next year. Called the *Anderson* she set the standard for many years.

In more recent years the Canadian hydrographic launch has evolved rapidly in a number of directions. The original gigs were of necessity open boats; the hydrographer had to have a clear line of sight for his sextant fixes. With the advent of electronic positioning equipment and the concomitant need to keep the equipment dry, cabins began to appear on the launches. Gasoline engines gave way gradually but inevitably to the safer and more reliable diesel. Wood was replaced first by fiberglass and later by aluminum. In fact even rubber or synthetic rubber is now being used in inflatable sounding boats that are carried on a ship's deck and powered by relatively small horsepower outboards. Outboard motors can be replaced with spares and the boat continues to sound while its motor is repaired; when a gasoline or diesel inboard breaks down, the launch is out of service.

The shape of sounding launch hulls has changed along with their fabric. The earliest launches used by CHS were double ended — sometimes derisively known as "canoes" — but today's boats sport a transom, broad or narrow depending on its maker. In cross section too the hull shape has undergone modifications. By varying the lines the designer can create what is known as either a "planing" or a "displacement" hull.

A planing hull carries the boat at high speed across the top of the water much as a hydrofoil does. This produces a speedy accumulation of sounding data for the hydrographer but it also produces a bone-jarring ride in choppy seas. On the other hand the displacement hull carries the launch lower in the water, with a less efficient consumption of fuel but a much smoother ride. Its disadvantage is dampness; in high seas the displacement hull ploughs through the waves instead of over them.

To meet the requirements of open-water surveying, accomplished over the course of days or weeks, the launch concept has been extended to boats of

Hydrographic Steamer, La Canadienne, *1906-1916.*

CGL Anderson *represented a 1942 evolutionary step in the development of CHS sounding launches. Previous designs had been smaller with shallow draft and were often forced to take shelter during even moderate blows.* Anderson *— and her descendants for years to come — was some 40 feet long, drew more than four feet of water and was powered with twin inboards.*

a size to provide accommodation for a reduced but, with the aid of technology, highly efficient ship's complement. Typical of this last type of vessel is the Central Region's 1973-built *Advent*, 23.5 metres long, 56 tons, with a cruising speed of nineteen knots (as opposed, for example to *Acadia*'s ten) and with only four crew to attend to the needs of three hydrographers (versus the *Wm. J. Stewart*'s fifty-two crew for eight hydrographers, and the *Acadia*'s sixty for ten). Stationed at the Canada Centre for Inland Waters (CCIW), Burlington, Ontario, the *Advent* is typical, along with the multipurpose ships used by both hydrographers and oceanographers, of the kind of vessel that will serve members of the Canadian Hydrographic Service in the future. Even the *Baffin*, the last of the big CHS ships, has just undergone (in 1982) a major refit so that it may better accommodate the needs of a variety of marine scientists.

But it is not only the ships of the hydrographic service that are changing. Vessels such as *Cartier, Acadia, William J. Stewart*, and *Baffin* demanded a certain type of sailing master and crew. Often a continuity of service was maintained also. Usually a number of years passed before a captain knew his ship with the degree of intimacy required for handling his vessel in frequently uncharted waters, often with confined sea room, and with the delicate manoeuvring necessary when launches were being hoisted or retrieved. The newer vessels, wonderfully engineered models of precision, lacking the sort of character (a euphemism for quirks, foibles, and downright failings) possessed by the older ships, ask less of their captains and crews in the way of forgiveness and long term loyalty. And so the men who sailed the old hydrographic ships have gone largely the way of their vessels.

The day is long since gone when men joined the service as seamen and, over the course of many years, worked their way up — literally as well as figuratively — from the fo'c's'le to the bridge. It was never a common occurrence but some sailors did sign on as deckhands aboard CHS ships and after a series of promotions find themselves installed in the masters' suites on the very same vessels.

One example was Captain Frank S. Green who spent his first day in the service of the CHS cleaning out the double bottoms of the *Wm. J. Stewart* after she arrived at Victoria, newly commissioned, in 1932. During his seagoing career he moved through the west coast hydrographic fleet — as master of *Parry* and *Marabell* — and eventually became captain aboard the same ship he had first signed on, going ashore for good only on his appointment, shortly before *Stewart* was decommissioned, as marine superintendent (deck).

Captain J.W.C. ''Jack'' Taylor lives today in retirement in Pictou, Nova Scotia. Between 1958-69 he was *Acadia*'s master. Captain Taylor is no sentimentalist but his voice — an instrument developed over years of barking orders from the bridge into the engineroom speaking tube, or to seamen below on the foredeck — modulates into something close to husky nostalgia when he recalls the loyalty of the men who served under him.

> They'd come back in the Spring [he recalls] say ''hello'' at the gangplank to the officer of the watch and go immediately below deck to their quarters. Each man, with his seabag and one suitcase, would go right to the bunk he'd had last season, unpack, and then start a conversation with his buddy, just as if they hadn't been apart from each other at all. They were all working together . . . shared the same job, sure . . . but it was more than that. They were *family*.

> Perhaps more nonsense has been written about the replacement of ships by other, supposedly cheaper means of working at sea, than about any other organizational problems in marine science; mostly the advocacy has come from aerospace and high technology industries who have a solution looking for a problem to solve. In fact it would be hard to find a working oceanographer or surveyor who would doubt that research ships will be their central tool for many decades to come.
>
> Report
> ''The Status of Research and Survey Fleet
> at Bedford Institute of Oceanography''
> September, 1982

The Chartmakers

The sad fact that hydrographers are little heard by legislators is reaching the stage of tragedy. These hard working and gifted men, working dutifully on grossly inadequate budgets, facing a task immensely greater than those faced by Hercules, are the only sane prop holding up the world from a total shambles in the ocean.

James Dawson
"Ancient Charts and Modern Mariners"
Marine Geodesy, Volume 4, Number 2, 1980

ONE HUNDRED YEARS IS A VENERABLE AGE FOR ANY HUMAN BEING and for most of mankind's institutions. The Canadian Hydrographic Service, which in 1983 celebrates the one hundredth year of its founding, is by no means the oldest governmental agency charting its nation's coastlines; the French Hydrographic Office was the first to be formed, in 1720 and the British hydrographic department started in 1795. The American forerunner of the present National Ocean Service followed five years later in 1800.

Though it wins no laurels for longevity in its century of existence and development, the CHS has achieved a distinction that is second to none. And, if one considers that age usually connotes experience, and experience wisdom, and wisdom a kind of placid easygoing temperament, one is hard pressed to explain the spirit of diffident aggressiveness that pervades the service from top to bottom. The paradox of a "diffident aggressiveness" can be resolved only by looking at two things. The first is a consideration of pressures imposed on the CHS from without, a condition which has led to an uncertainty among hydrographers about who they are, and where exactly they belong in the structure of governmental service.

The second consideration requires looking at a few of the men — and, in latter years, the women — of the service who through their work and their attitudes have endowed Canadian hydrography with a tough, no nonsense approach.

The first influence upon the CHS was, and continues to be, uncertainty at the highest levels of government about where Canadian hydrography belongs. Since 1883, one hundred years ago when Boulton drew his assignment to chart the waters of Georgian Bay, the CHS has been shifted from ministry to ministry no less than fourteen times. That is, the responsibility for administering the CHS has changed on the average every seven years.

Examples of this governmental juggling are rife but one will suffice. In June 1936 the government formed a new Department of Transport and it was proposed to shift the CHS to it from the Department of Marine where it had been operating since 1930. The Dominion Hydrographer, Captain Frederick Anderson, objected strenously to the move in a letter to the deputy minister.

The CHS, Anderson wrote, ". . .is not a survey organization; it is a marine service . . . I consider it my duty to point out . . . that this proposed measure whereby a distinctive nautical service would be pulled up by the roots . . . and planted . . . where operations and activities are vastly different would not be in the best interests of the public it serves." His plea went unheeded.

Allied with this governmental tendency to move the CHS from department to department is the almost reclusive nature of hydrography itself. Hydrographers have been compared to physicians engaged in preventive medicine. Physicians and other scientists practising preventive medicine work to head off epidemics, to prevent diseases from occurring. Their greatest success is the lack of illness — *no* plague, *no* epidemic, *no* public panic. In like fashion, hydrography's task is the prevention of maritime disasters and, as with preventive medicine, hydrography's greatest success occurs when no ships sink, no lives are lost at sea, no oil spills occur along our shorelines.

To an astonishing degree Canadian hydrographers have been successful; as the nation with the world's longest coastline we have been fortunate in suffering as few groundings as we have. But this kind of success, marked by a lack rather than a surfeit of publicity, breeds anonymity. In turn anonymity can lead to neglect and the possibility that essential though unheralded services are underfunded. The development of Salk vaccine for infantile poliomyelitis may have been delayed for years had not the outbreak of disease in the 1950s prompted massive spending by the United States' government to find a preventative. Had the SS *Asia* not foundered in 1882 in Georgian Bay the establishment of a government hydrographic survey may well have been delayed.†

Together, these official tendencies either to ignore Canadian hydrography when it's functioning efficiently, or overreact when it is seen, correctly or not, to be inefficient, have produced in the men and women of the CHS a kind of diffidence. They tend to lay back, to be restrained, reluctant to boast or brag about their accomplishments.

The second influence — and, again, a continuing factor — on the CHS has been the quality of people who have enlisted in its ranks. From the day in 1884 when Boulton recruited young William James Stewart, freshly minted as an engineer and gold medallist from the Royal Military College at Kingston, the CHS has attracted tough-minded individualists, usually as strong in body as they have been in determination and supremely confident in their ability to provide detailed, accurate charts, the equal of the best produced anywhere.

In this chapter, we take a look at some of the personnel of CHS, the staff that ventures out from the four regional offices of the service to plot the coastlines of the country's three oceans and innumerable inland waterways, and to sound the depths of all waters.

Henri Delpé Parizeau was seldom modest about his work and indeed he had little reason to be. He was a hydrographer and, in common with most

†It is an irony of fate that although *Asia* did not sink from a lack of adequate charts her demise did lead to the establishment of the CHS. At the time it was suggested — largely by the ship's owners and her insurers — that *Asia* had foundered upon an uncharted shoal off Parry Sound in Georgian Bay. The subsequent court of enquiry ruled the ship had gone down for three reasons, none of them connected with uncharted shoals. The reasons: the ship was ill-designed for the waters in which she sailed; she was top-heavy and overloaded; she was struck by a freak storm of almost hurricane proportions.

"Feeding" a stricken ship

The hydrographic ship, *Chrissey Thomey*, was severely damaged by ice floes on a surveying expedition in Hudson Strait in 1911. She was saved from foundering by a technique called feeding. An old sail was slung under her bow at the point where her timbers had been stove in. Several tons of ashes (from the hold of the steamer *Minto* which was her escort vessel) were dumped into the sail. Water rushing into the *Thomey*'s hull carried the ashes into the breach and sealed it off. Though she had taken more than five feet of water into her hold, the temporary patch held while she proceeded under her own power to a sheltered cove. There the schooner was beached and repaired.

of his breed, a pragmatic realist. There was no room in his professional life for false modesty, for the "oh, it's nothing" disclaimer. So assertive and even truculent did Parizeau become that it has been suggested that his eventual appointment as regional hydrographer in British Columbia was prompted, in part, at least, by William J. Stewart's desire — he was then head of the service — to remove his undeniably talented but difficult assistant to as distant a posting as possible.

Eddies of controversy and dissension swirled about Parizeau but he was a kindly though demanding employer. R.B. Young, the hydrographer who was later to be Pacific regional chief, has recalled the experience of working for him.

"I joined the service in 1929 fresh out of engineering school," Young says. "I'd had some experience in land surveying but none at sea. Mr Parizeau was a considerate and thorough instructor. Once we learned what it was we were to do, he expected the highest quality of work from us; his standards were very high. But if his staff was ever abused or criticized he defended us to the hilt.

"I guess you'd have to say he was an excellent man to work for but I think he might have been a bit difficult to manage as an employee."

Parizeau would have been greatly amused at the words quoted at the beginning of this chapter comparing hydrographers to the Greek hero Hercules. Nonetheless Parizeau did perform herculean feats in the service of Canadian hydrography; he did have his works ignored by the legislators; and his was a loud — if sometimes strident — voice for sanity on the oceans.

Henri Parizeau was a short, dapper, rotund Montrealer, the eighth son of a well-to-do Quebec lumber king. With some of his older siblings firmly settled into the family business Henri chose engineering as a career and took his degree at McGill. For a few years he worked in private industry but by 1901 had joined the federal government service. In 1906 he was named as senior assistant to the Pacific coast survey, then newly established.

Parizeau first began surveying the coast around the mouth of the Skeena River where Prince Rupert was being developed as the western terminus of the Grand Trunk Pacific Railway. But in 1910 he was recalled to hydrographic headquarters in Ottawa.

At the time, the Dominion contemplated building a railway from Winnipeg to the shores of Hudson Bay as a route for prairie grain to be shipped to Atlantic and European ports. The ships would require a deepwater port on Manitoba's northern shores. Instructions from the Department of Marine and Fisheries (under which the hydrographic service operated at the time — 1909) stated that the chief hydrographer was to "make out a report . . . as to what steps should be taken for a survey of the approaches to either Port Nelson or Fort Churchill." Stewart, the chief, named Parizeau to conduct the Port Nelson survey, Charles Savary to Fort Churchill.

A sand sucker, used to dredge the approaches to Port Nelson on Hudson Bay, crashes into the dock and is demolished during a 1914 storm. Henri Parizeau's prediction that Nelson would prove inadequate as a port was confirmed. Eventually Fort Churchill was chosen as the terminal through which prairie grain would be shipped.

Parizeau sailed north in 1911 in the newly purchased and refitted three-masted schooner *Chrissie Thomey*. The sailing master was Captain Thomas Gushue from Brigus, Newfoundland. He was known to his crew as Black Tom O'Brigus, more for the colour of his hair and beard than his temperament. In two survey seasons Parizeau, with the help of but two assistants, completed the survey of the port approaches. Port Nelson was not the ideal spot for a saltwater port, he concluded. In his annual report of 1912 Stewart repeated Parizeau's caution: "Port Nelson is difficult to approach and hard to pick up. This may be remedied by light ships and gas buoys and the creation of a town, but can never rival the easy access of Churchill."

For reasons of its own† the government chose to ignore Parizeau's advice. Work proceeded at Port Nelson; the very next year disaster struck.

The year 1913 was CSS *Acadia's* first season and as mentioned in the previous chapter she was the spanking new darling of the service. On 12 October while lying off Port Nelson, *Acadia* was caught in a 60-mile-per-hour gale that swept her new 34-foot gasoline launch off her deck into the sea. *Acadia* her-

†It is impossible — and unnecessary in a book of this nature — to sort out at this late date the government's reasons for ignoring Parizeau's advice. Perhaps the fact that Nelson lay approximately one hundred and fifty miles south of Churchill and that much closer to Winnipeg was a consideration as the cost of laying track would be considerably less. Another factor may have been that during the time of Parizeau's surveys responsibility for hydrographic services was transferred from the Department of Marine and Fisheries to the newly created Department of Naval Services; what appears at a distance of nearly three quarters of a century to be deliberate oversight may have been nothing more than a bobbling of bureaucratic responsibility.

self, with both anchors out, was dragged for about a mile and a half before the anchor flukes bit and held; on her maiden voyage north, the pride of the service came close to destruction.

One week later the supply steamer *Alette* went down in heavy ice, again off Port Nelson; *Acadia* picked up the twenty-eight crewmen. On the return passage the presence of the extra men aboard put a strain on provisions; the ship arrived back in Halifax with hydrographers, crew and wreck survivors gaunt from short rations.

Two more ships went down, and a million dollar dredge — a sand sucker, actually — rammed into the dock at ill-fated Port Nelson before World War I interrupted all work on the projected Hudson Bay port. It was 1928 before the Dominion government again began work on a northern port and by that time the decision settled on Fort Churchill — as Parizeau and Stewart had recommended fourteen years before.

Perhaps he spoke a little too soon

The charting begun in Hudson Bay by Parizeau and Savary, and continued after 1928 by many others, was effective. As Meehan reports "there have been few major ship fatalities along the Hudson Bay Route, and these were not for want of adequate chart coverage." Captain W. Mouat of Newcastle, England, took the grain carrier SS *Pennyworth* into Churchill in 1932 and later reported that "as this vessel was fitted out with a gyro compass and echometer we experienced no trouble whatever the courses being maintained as set . . . navigation in these waters is far easier than navigating the St. Lawrence." He might have been predicting his own fate; the very next year Captain Mouat managed to run *Pennyworth* aground on a rock in the Saint Lawrence below Quebec City.

At another time, in another place, the presence of S.R. "Steve" Titus loomed on the Canadian hydrographic scene. Titus was a giant in CHS, a giant in at least two senses of the word; by those who worked for and with him, he is variously known as the "father of the modern CHS", the "man who ran the service", or the "autocrat of Ottawa". Second, he was literally a giant — "I never knew his actual height," says one former colleague, "but he stood six foot five or six" and drove a big Cadillac car, "twenty-five fathoms long with a permanent list to port."

Titus was never known to button a shirt about his neck or wrists; the broadcloth simply wouldn't meet. His hands were massive appendages and more than one young hydrographer wondered how Titus ever managed to hold a sextant without crushing it.

In 1948 Steve Titus was appointed hydrographer-in-charge aboard *Acadia*. When he lumbered into his stateroom for the first time he looked into the bathroom and snorted in derision; the bathtub that had served all previous hydrographers-in-charge for thirty-five years appeared to be nothing more than a foot bath to him. "Replace it," he ordered.

The oversized replacement served Titus for two years, and his successor as *Acadia's* chief hydrographer, Colin Martin, a man much closer to statistical averages in physical size probably enjoyed the extra space the tub provided. Martin's successor, however, was Hiro Furuya who stood a foot shorter than Titus and weighed about one hundred and fifty pounds. Titus's bathtub became known in the service as "Furuya's swimming pool".

One could never ignore Titus's powerful physical presence and this probably helped in the role he chose to play when he was eventually promoted to a desk job at headquarters as supervisor of field surveys. "He was a dictator," a former colleague recalls, ". . . benevolent, but a dictator nonetheless. He had a vision of what the CHS should be and, by God, everybody had to *work* to attain that vision."

In spite of his autocratic methods Titus was a kindly employer. Throughout his career he remained a bachelor, living in Ottawa with his mother during the latter years of his service. Only after his retirement did Titus marry.

"He used to invite us young guys over to his house to spend an evening," says Ross Douglas who worked with Titus before his retirement. "We'd sit around and talk hydrography or watch television. Steve's favourite snack during these sessions was a pineapple — a whole pineapple. He'd sit in his huge armchair and carefully peel the fruit with his jacknife. Then, he'd carve it into slices and pop them — whole — into his mouth."

. . . to the crew, hydrographical surveying is extremely monotonous work, which is the hardest kind of all work — that, unlike seafaring men in other employs . . . they do not see their homes from the beginning to the end of the season. . . . That giving the depth of the water — a mistake in which may involve the loss of a vessel and lives — is a responsible duty, and can only be entrusted to intelligent and trustworthy men. . . . in order to induce first-class men to join and stay in this work, good wages must be given, and every material comfort that the circumstances can afford.

J.G. Boulton
Staff Commander, Royal Navy, in
charge of Georgian Bay Survey
in his annual report, 1884

The alarm clock jangled tinnily and brought Chris Rozon slowly, reluctantly, out of a deep sleep. It was 0645 hours. Chris withdrew an arm from the warmth of the blanket and slapped the clock into silence. The arm was tired, ached dully in every muscle. But then, so did just about every other muscle in Chris Rozon's body. Shoulders stiff and reluctant to move; hips and thighs tender and mottled with bruises that ranged in colour from fading yellow and brown of last week's collisions with the launch's gunwhales to the startling purples of yesterday's.

For twelve hours yesterday from 0800 to 2000 hours Chris had been aboard CSS *Baffin's* launch *Finch* running sounding lines in Fury and Hecla Strait in

"Mugging up."

the Arctic. The sea had not been particularly rough — nothing more than two- to four-foot swells — but at a speed of nine knots *Finch's* hull had pounded into the waves with jackhammer force. Chris, along with the launch's coxswain, and its seaman had been bounced about like popcorn in a frying pan.

Twelve hours of work — most of it hard, physically punishing labour — and Chris Rozon woke to another day of the same routine. To her own amusement she found herself looking forward to it with anticipation.

Christine Rozon is a hydrographer with CHS. It would be an absurdity to typify her as "average" in anything; still she is representative of the young people who today join CHS and pursue careers in the tradition of Cook, Vancouver and Boulton.

Today as in the past, most field hydrographers with CHS begin as either graduate engineers or surveyors. Chris Rozon is an exception; she graduated in 1978 from Mount Allison University in Sackville, New Brunswick, as a marine biologist. In 1979 she was hired by the Bedford Institute of Oceanogrpahy (BIO) as a researcher; the institute was conducting surveys — hydrographic and statistical — into the possible effects of the proposed tidal hydroelectric power installation at Passamaquoddy Bay. Chris's job was to study a small marine organism that lives in the water around the site of the development and determine what effect the power plant would have on its life cycle.

"I was hired on contract [non-permanent staff] and spent the summer at the institute hunched over a microscope studying these little beasts," Chris says. "But all during that time I was spending my spare moments — coffee breaks, lunch hours — with the guys who were hydrographers. I had always been attracted to the outdoors — skiing, hiking, and that kind of thing. And the hydrographers' description of the work they did — out of doors, all of it — sounded much more exciting and gratifying than sitting in a lab. The next year I applied for a position as a hydrographer and was taken on."

Alex Raymond's route to the CHS was more typical. Graduating from a technical institute with a diploma in surveying he joined the service in 1971. In the years since, he has served aboard all of the Pacific Region's ships on surveys ranging from British Columbia coastal inlets to the western Arctic. He's acquired considerable experience and expertise, a wife, Cathy, and a young son, Ross.

Till rigor mortis do us part

Hydrography at its best requires team effort. Everyone from the stewards and cooks aboard ship, to mechanics and electronic technicians in the workshops, to the coxswains and seamen who man the launches, plays his unique role; woe betide the team member who shirks or bungles his part of the operation. Though every member of the team is important, the linchpin, from the hydrographer's viewpoint, is the coxswain. It's the coxswain's responsibility to steer the launch and to steer it accurately by compass or black box; if he veers off a sounding line it must be run again to ensure bottom coverage. Naturally a close relationship builds up between the coxswain and hydrographer; each has his favourite partner and each pair strains to achieve the ship's best record in terms of work done accurately. In this competitive atmosphere few incompetents survive. One did, for a time, and it was hydrographer R.W. Sandilands' misfortune to encounter him aboard a sounding launch prior to the day's run. "He was introduced to me as Rigger," Sandilands says, "or so it sounded. Within a few days I was struck by the inappropriateness of the man's name. He was a menace afloat, couldn't operate the boat properly and I seriously doubted that he could rig, splice or perform any seamanlike task. Later, I found out that I had heard his name — or nickname — correctly. But I had mistaken the spelling — it was Rigor. His mates had dubbed him Rigor Mortis in recognition of his total inability to function."

Acadia's mercy missions

In everything he does the hydrographer is concerned with safety — the safety of ships, crews, passengers, cargoes. Sometimes, however, his work places him in a position where he is more directly involved in saving lives than when he is producing accurate charts. In the late summer of 1961 CSS *Acadia* was twice pressed into service to rescue Newfoundlanders trapped in outport communities threatened by forest fires. "On 7 August I got a radio message from the deputy minister of the Newfoundland Department of Welfare," says Hiro Furuya, the hydrographer-in-charge then, later chief of training and standards at headquarters. "He asked me to keep the ship on stand-by at Musgrave Harbour [near the northeastern tip of the island] in case the residents had to be evacuated." The next day 298 people — men, women, children and eight stretcher cases — came aboard in small boats and were evacuated to Seldom-come-by. On 10 August Furuya and the ship's master, Captain Jack Taylor, picked up an additional 204 stranded people from two smaller communities.

Now, in mid-career, Raymond has his mind firmly fixed on promotion to hydrographer-in-charge of a major survey ship. He recognizes that his ambition carries a price tag, a steep price indeed. Half of each year while he's at sea, the full burden of running the Raymond household falls on Cathy. Then, too, he is absent for many of those occasions — his son's first step, first word, first day at school — that most parents witness and cherish.

As has been recounted, the CHS began in 1883 with the establishment of the Georgian Bay Survey. Lacking experienced men of its own the Dominion government accepted the offer of the Royal Navy to provide the services of Commander Boulton. Boulton himself was both a mariner and a hydrographer and part of his commission in establishing the Canadian service was to recruit Canadian nationals to the survey. At the time Canada had no backlog of trained experienced mariners; expert offshore fishermen existed in quantity on both Atlantic and Pacific coasts but they were, by and large, sailors who navigated by "feel" — that is, by instinct, intuition and experience. Boulton recognized that he needed men who possessed more, who were trained or trainable in the intricacies of mathematics, astronomy, and even possessing some talent in draftsmanship. He opted for trained surveyors without qualifications as mariners and his first recruit to the service — William J. Stewart — filled Boulton's requirements exactly.

Stewart's enlistment in Boulton's survey confirmed the pattern followed ever since by the CHS — the survey and the surveyor are paramount.

Indeed so firm was this rule that until the late 1950s the hydrographer-in-charge aboard any CHS survey ship was in full command, even when the ship was underway. The hydrographer was in *complete* charge — he ordered stores, and served as paymaster for the entire ship's complement. The hydrographer-in-charge was considered inferior only to God, and aboard ship his word prevailed.

Although the title was never conferred officially, some hydrographers-in-charge became known aboard their ships as "commander". Captain Jack Taylor, who was the last master aboard *Acadia* before she was decommissioned, remembers one of his early voyages on *Acadia* when he was approached on the bridge by the radioman on the first day of the new survey season. The man asked the captain if he knew where the "commander" could be found. While he occupied the bridge as a newly minted master, Taylor had long years behind him as mate and seaman; he knew the hierarchy of the sea and he knew "commander" signified, in the radio officer's mind, an officer superior to the captain. Taylor sharply corrected the man's terminology, directed the radioman to the hydrographer-in-charge and never again heard the word used on his ship.

Such annoyance must have been common in the early days of Canadian hydrography. Even today the relationship between the ship's master and hydrographer is a delicate one requiring understanding on both parts. The reasons for potential conflict between master and hydrographer are many but can be stated briefly in this way: the master's prime consideration is for the safety of his ship and the men who sail in her; generally this means that the master will want to keep as much water under his keel as possible. The hydrographer, on the other hand, is interested in taking the ship — or the launch — as close to a reef, shoal or rock as he can, the better to obtain accurate figures on the water depth over it.

The ship's captain may be just as interested in getting accurate soundings as is the hydrographer; the hydrographer is certainly no less interested than the captain in the ship's safety. Yet the demands of their two jobs may put the men at odds.†

Not unnaturally this contentious issue has led to many clashes of personality. In the end though, common sense and a common regard for the good of the survey have prevailed; there have been only minor instances of grounding — a hazard common to all survey ships — or collision due to this split command. The only major instance occurred in 1957 with the newly commissioned *Baffin*.

Captain J. W. C. Taylor now lives in retirement in Pictou, Nova Scotia, from which port he sailed for several years as Acadia's *last master before she was decommissioned.*

†The hydrographic services of almost all other maritime nations — there may be one or two exceptions in addition to Canada — require hydrographers in charge of survey ships to be masters as well; that is to say that the officer in charge of the survey ship is both hydrographer-in-charge and captain. Within CHS there are, in the mid-1980s, several hydrographers — perhaps a dozen or more — who also hold their master's certificates but the Canadian tradition has been to separate the functions.

The last sailing gig used for sounding, 1931.

Baffin was on a "shakedown cruise", prior to proceeding north on her maiden voyage to the Arctic, but the trip ended ignominiously. The ship was surveying off Cape La Have, west of Halifax on Nova Scotia's south shore. As related earlier, *Baffin* was testing the relatively new Decca two-range system of electronic positioning. For the two previous seasons on CSS *Fort Frances* and CSS *Kapuskasing*, the plotting had been carried out adjacent to the bridge. On *Baffin*, it was done in the drawing office one deck below. Almost unbelievably there was no echo sounder on the ship's bridge, but in the drawing office only.

At 1630 hours on the afternoon of 4 July 1957, CSS *Baffin* was approaching the inner end of a sounding line in thick fog but at her normal sounding speed. As was the practice then, the senior hydrographer on duty was giving orders directly to the helmsman. Shortly after 1600 hours, the first officer identified Black Rock on the ship's radar when it was a little less than two miles away. He informed the hydrographers when it was one and one half miles away and twice more.

The hydrographers preferred to believe their Decca readings. What they did not know, and what became clear only much later after a detailed analysis of the entire day's notes, was that during a rainstorm that morning, the Decca system had "lane slipped" and the Decca position was in error by more than a mile. A last minute change of course put *Baffin* right on Black Rock.

It took five days to refloat the *Baffin* and she spent so much time in drydock repairing the damage that she did not go north that season.

A judicial enquiry, under the Canada Shipping Act, was held in February 1958. The most important part of the decision read: "The court believes that over the years there had developed a system whereby there was divided responsibility between the Captain and the hydrographers regarding the navigation of hydrographic ships . . . The Court finds that the primary responsibility for the direction and position of the 'Baffin' during survey operations was one assumed by the hydrographers both on account of practice and the fact that they were in the position of owners of the ship. . . . There was, however, an additional undefined ultimate responsibility on the Master . . . although this seems to have been confused over the years."

The report concluded:

> . . .The Court doubts if any person or persons can be blamed specifically for these unfortunate events. The "Baffin" was a new ship and might be considered a victim of history in that over the years there had developed a system which was not workable aboard this ship. There was lack of understanding as to responsibility and too much confidence placed in the equipment. The systems of the past were no longer applicable and it was unfortunate that this was not realized at an earlier date. The ship was going too fast under the weather conditions . . . No one seemed to realize the dangers and the possible difficulty. There was an innocence about the whole matter which is difficult to understand. . . . The Court with the reservations already indicated finds no intentional wrongful act or default on the part . . . any . . . persons on board the "Baffin" which caused or contributed to the grounding. . . .

Life in the Canadian Hydrographic Service would never be quite the same again. From then on the master of a hydrographic ship was undoubtedly responsible for safety of the ship at all times. Since then Canadian hydrographers have been privileged but restricted passengers aboard ships that carry them on the surveys they direct.

In the minds of some of the service's older hands this "demotion" may still rankle. It's a measure of these men, however, that no matter how much they yearn for the "good old days" when they virtually assumed command of their ship, they put these thoughts behind when the requirements of the survey become paramount.

But putting the ship under the captain's full command did not prevent further accidents. In the summer of 1973 George Yeaton, now chief of nautical geodesy at Ottawa headquarters, was hydrographer-in-charge aboard the chartered ship MV *Minna* surveying off the northern coast of Labrador.

"It was a multidisciplinary survey," Yeaton recalls, "hydrography, seismology, gravimetry, magnetometry — we were to produce what are known as 'natural resources charts'. We'd been at sea for four or five weeks, our supplies were getting low and the whole crew, hydrographers and seamen, needed some rest and recreation. We put into Godthaab, Greenland, to pick up supplies and re-acquire our shore legs."

With the ship resupplied and her personnel refreshed, *Minna* sailed for Resolution Island where Yeaton had deposited a crew of technicians to build

Lifeboat drill.

a Decca tower. The plan was to pick up the crew and return to running survey lines.

"We started into a small bay that had never been properly charted but we knew it was deep water," Yeaton says. "Even so, I didn't think we should get too close to shore. I went up to the bridge to tell the captain that since it was blowing pretty hard I thought the best plan might be to drop anchor and I'd send a launch ashore for the men.

"*Minna* was a large ship — some sixty-five or seventy metres — but she didn't have twin screws like *Baffin* — which was about the same size — just one. So she wasn't very manoeuvrable, especially in rough weather. The captain said 'no', he'd take her in. As I say, it was blowing pretty hard and the ship was slewing about. As we approached the shore the bow hit a pinnacle and — damn! — she was fast aground."

Like many deepwater vessels, *Minna* was double bottomed and the rock penetrated only her outer shell; but the pinnacle projected well into the space between the double bottoms and held her fast. "With the wind and the waves she started to work back and forth and it began to look like we'd lose all of our equipment," Yeaton says. "Next morning I organized the whole crew and we stripped as much of the electronic gear as we could get off."

Yeaton radioed his office at BIO, Dartmouth, for instructions and was told a DOT icebreaker was in the area; Bedford would request help and he could expect some kind of response soon. Within days CCGS *N.B. McLean* steamed into the bay, put a line aboard *Minna* but couldn't budge her.

"Then we heard that an RCN flotilla was in the area on exercises. Once again Bedford made the arrangements and five or six destroyers and a large supply/repair ship, the HMCS *Provider*, steamed into the bay." The navy organized removal of the scientific equipment from the beach — "in half a day with helicopters and rubber boats they evacuated about a million dollars' worth of equipment."

Yeaton, his hydrographers and technicians were eventually flown out to Frobisher Bay where Ken Williams, hydrographer-in-charge of *Baffin*, arranged to pick them up and return them to Dartmouth. *Minna* was abandoned and eventually sank. Before he had removed his sounding equipment from the

Bear facts of the Arctic

Arctic surveys present hydrographers with problems not usually encountered elsewhere. Not the least of these problems is the one presented by the wildlife. Most of the encounters are with curious but shy creatures. Bears are the exception. On one survey in the western Arctic a lone polar bear became unnaturally fascinated by a radio transmitter installed by the survey crew. It prowled about the equipment, sniffing at it and occasionally buffeting it with its massive paws. Its continued attention might have delayed the party's work or destroyed the equipment but the men could find no way to discourage the creature. Finally it was discovered that a liberal sprinkling of mothballs around the base of the transmitter was sufficient to distract the bear.

On another occasion hydrographer Barry Macdonald complained that in all of his considerable experience on Mackenzie River surveys he had never seen a bear. His companion, Phil Corkum, answered: "No? Well hie yourself down to the camp outhouse. At this very moment it's occupied by one." A gag, of course, thought Macdonald but he made the short trek down to the privy to discover that a largish brown bear, festooned with toilet paper, was indeed occupying the convenience.

wreck Yeaton had made one final sounding; the bow of the ship was high, held firm by the pinnacle that pierced her hull. A mere ten feet aft Yeaton couldn't find bottom with his leadline. "So I turned on the echo sounder", he says. "It registered twenty-five fathoms — one hundred and fifty feet."

Fred Smithers was a young graduate draftsman when he joined CHS in 1941 at the Pacific regional office in Victoria. Trained to work at the drawing-board, Smithers found himself at sea — literally — within weeks of signing on. "It was wartime, of course," he says. "Many qualified surveyors had enlisted and simply weren't available for civilian work. So Mr [Henri] Parizeau [the regional hydrographer at the time] pressed into service anybody in the office who was able to swing a leadline and hold a sextant." Smithers himself, when he was old enough, tried to enlist "but Mr Parizeau effectively blocked that."

While the war lasted, Smithers surveyed portions of the British Columbia coast from the houseboat *Pender*; much of the work involved surveying sites for RCAF Coastal Command stations from which the flying boats were operated. "It was six months on the water, and then six months back on shore drawing the charts," he says.

It was still the pre-modern era when the same person took soundings and positions, and later transferred them to the master sheet that was sent to Ottawa for engraving. "We did everything — *everything* — by hand," Smithers says. "We converted every figure entered on the field sheet to the chart scale — by hand . . . no calculators, no computers. We drew the shorelines, shoals, and every feature — by hand. Then, when we had the chart drawn, we lettered the whole thing — by hand. In the early years I used to spend hours almost every evening with a textbook of letterforms in front of me practising the different typefaces, getting the serifs right, making certain the letters lined up properly. Exacting work but it was fun."

When the war ended and returning servicemen trained in surveying again became available in numbers, Smithers went ashore and became — what he thought from his enlistment he would be — a cartographer rather than field hydrographer.

It had been traditional in the Royal Navy that the hydrographer-in-charge of any survey signed the charts he produced. Usually this was done in the form of a legend in the chart's title reading something like: "Based on surveys conducted in 1875-7 by Capt. John Doe, R.N. with assistants So and So and So." The tradition was taken over intact when the Canadian service began. It was appropriate enough when the field hydrographer was the same man who actually drew the chart, which was the system when Fred Smithers began his career. But when the increasing demand for charts, and the increasing complexity of the task of providing them forced the separation of hydrography and cartography, many of those who contributed to the production of the charts went without published credit. The hydrographers still had their names displayed in the chart's title block but the cartographers did not.

"We felt we had some claim to credit, to 'immortality'," Fred Smithers says. "On each chart we produced we'd work our initials into it somewhere. Usually it was along a shoreline where we used a lot of squiggly lines to indicate rocks at low tide. I'm pretty sure Mr Parizeau knew we were doing it but never caught us at it or, at least, never ordered us to stop."

Today, the practice of crediting the hydrographers in the chart's title has gone the way of many another tradition; no chart based on contemporary surveys lists the name of any individual, hydrographer or cartographer. Indeed, Fred Smithers's ingenuity in concealing his initials is made impossible with the use now of pre-printed, adhesive overlays to indicate hydrographic and topographic detail.

Hydrographers old enough in the service to remember the "good old days" mourn the loss of their names on the charts they've helped produce. They consider the charts to be their "professional papers", analogous to the papers published by research scientists after their field or laboratory work is done.

In the one hundred years that the Canadian Hydrographic Service has been in existence thousands of men and women have served in its ranks. Some have served but a short time since the turnover of field staff is and always has been high. Absence from home and family for five to six months of every year

A tale that sheds no light at all

In the early years of the twentieth century Canadian hydrographers frequently used lighthouses as principal points in their triangulation networks. Often enough the surveyors were the lightkeeper's sole visitors during navigation season. One year hydrographers aboard *Bayfield* discovered that the light near Saint Ignace Island on the north shore of Lake Superior was not operating. Putting ashore they found that the keeper and his wife, experimenting with diluted wood alcohol as a beverage, had managed to kill themselves with alcohol poisoning. William Baker, second engineer of the ship who was described as an "ingenious and likable — but somewhat irresponsible — character", maintained that the light was operated for the rest of the season by sticking a wick in the woman's mouth, igniting it, and placing her within the lantern. He claimed that the light burned until the close of navigation.

imposes a strain that many do not consider worthwhile.

Others in the service have stayed for years, many of them serving their whole careers with the CHS. Obviously with such a diversity of service and such a variety of personalities it is impossible to generalize to any great degree about CHS personnel, past or present. Some broad characteristics however mark all of them.

As noted, hydrographers tend to be individualists; indeed it might be said that any person who is not strong in a sense of his own worth would be lost in the profession. Hydrographers in the field are usually part of a team but almost always a small team; they manage the field work in parties frequently numbering no more than three or four persons. But even within the group each member is frequently left to himself to accomplish a specific task. It takes a sturdy sense of self-confidence to function in isolation.

Then, every hydrographer is either born with, or acquires, a compulsion for accuracy. In Canadian hydrographic circles it is common to hear the phrase "nit-picking" but it isn't used in a derogatory sense. Hydrographers know or are taught (and if they don't learn they cease being hydrographers) that the essence of their job is accuracy. With the one exception of air traffic controllers it is difficult to think of any occupation on which the lives of so many people depend as on that of the hydrographer. And to say that much is to ignore the safe delivery of millions of dollars of cargoes, the safe passage of billions of dollars worth of shipping. So hydrographers are nit-pickers and proud of it.

It's the shared work, of course, that primarily unites the men and women of the CHS into a closely knit fraternity. But there's more. The dairy farmer running a Holstein herd at Lawrencetown, Nova Scotia, may have much in common with the Woodstock, Ontario, farmer and his Jersey herd; but once they've discussed the idiosyncracies of the milk marketing board, and the intricacies of vitamin feed-additives, they've more or less run the gamut. Not so with hydrographers. The geography of Canada may separate them by hundreds or thousands of miles but there are factors *other* than the work itself that unite them.

The first factor is training. Within a year of being hired, a CHS hydrographer must take — and pass — Hydrography I. It is a five-month study program, half spent in a classroom at Ottawa headquarters, half in the field on survey. After another four years the hydrographer — if he or she wishes to advance beyond the working level — must take Hydrography II, an advanced course. No matter to which of the regions the hydrographer is eventually assigned, he or she finds his associates have shared the same educational experience and can reminisce about the quirky habits of the same instructors.

At another level is the university training plan within CHS – UTP as it's known. A hydrographer with five or more years in the service, who is highly motivated, qualified academically, and who has demonstrated the necessary leadership potential is invited to apply for university training at the service's expense. It may be that the hydrographer is already a university graduate; no matter. If he or she shows managerial ability, is geographically mobile, wants to increase his/her educational level in an area considered useful to the service, the service will send him/her back to college. During vacations the hydrographer works for CHS on surveys or in the office.

Ross Douglas, now regional director of hydrography at CCIW, Burlington, recalls a telephone call in 1975 from G.N. Ewing, then Dominion Hydrographer. "I'd been a hydrographer since 1960," Douglas says, "but Gerry said, 'Douglas, I'm going to give you one last chance to go to school.' He didn't care whether I took psychology or nursing, he said, but it was bloody well my last chance."

Douglas did return to school — and graduated from Dalhousie in geology, a science that had fascinated him since his earlier years in the Arctic as hydrographer with the Polar Continental Shelf Project (PCSP).

Another experience all hydrographers share is the tri-yearly appearance before the appraisal board. The Dominion Hydrographer and each of his four regional directors sit in judgement on every hydrographer's performance, assess his plans for the future and advise on career advancement. Some hydrographers consider the assessment hearings a challenge, while others liken them to the Inquisition; all of them share the experience, and it binds them together as would passing through a fiery furnace hand in hand.

Finally, there's the annual Canadian hydrographic conference, the twentieth in the series being held in April of the centennial year. A hydrogra-

It pays to patronize the local pub

In the spring of 1960 four graduating students of a Calgary surveying school met in a tavern on 16th Avenue to toast the end of exams. None of them had lined up a permanent job and the conversation naturally involved their prospects. At one point in the conversation, one of the quartet pulled from his pocket a poster announcing a job competition for surveyors' positions with the CHS. None had the slightest notion of what hydrography was and the word "shoal" baffled them; they applied nonetheless. D'Arcy Charles journeyed west to interview them and hired all four. Three remained with the CHS. They were: Neil Anderson, who went on to become director of planning and development at Ottawa; Ross Douglas, regional director of hydrography at CCIW, Burlington; and Earl Brown, assistant regional hydrographer, CCIW.

phic conference is like none other. The 1983 affair, with the theme "From Leadline to Laser", was an intense three-day session of workshops and seminars. As its title indicates it covered hydrography from the past to the future. But after the daily schedule was concluded, the hydrographers socialized. "There's just as much . . . or more . . . comes out of the time spent around the tables in the pub as around the tables in the conference room," says one participant. In fact, at one conference of recent years, the host hotel ran out of beer several hours before closing time; emergency supplies were rushed in.

Though they are brought together by shared work and experiences into what is something like a lodge with secret signs and handclasps, Canadian hydrographers are — largely — an anonymous group in public. As said earlier it is usually only when a marine disaster occurs that attention focuses on their work. Hydrographers have reversed the show business adage that "any publicity is good publicity" and have tended to advocate that "no publicity is best".

The Canadian hydrographer was in 1883, and his successor remains in 1983, a sturdy individualist pursuing his unceasing search for accuracy in a largely anonymous profession. Nonetheless it is still possible to see some difference between those who practised the craft one hundred years ago and those who engage in the science today. As in every aspect of hydrography, technology has produced the greatest change in those who practise it.

From the earliest days of Boulton and Stewart through to the 1950s and 1960s Canadian hydrographers were rugged outdoor men with all that implies; it was a mere accident of Canadian geography and weather that forced them to spend half of each year indoors. Granted, the season ashore permitted them time to transfer their field data to a form — charts — in which they could be used by mariners. It was nevertheless the field season, either afloat or in shore parties, that justified their existence.

Today the field season is no less important; it is and always will be the essential first step before a chart can be issued. But the men and women who conduct the surveys in the field have changed in other ways.

The change had been coming for some time but if a turning can be pinpointed it was the arrival of Doctor W.E. van Steenburgh as Director General in the Department of Mines and Technical Surveys, then "home" to the CHS.

Van Steenburgh brought a scientific mind and the scientific method to his department. It was van Steenburgh's vision that established the PCSP in 1958. He brought hydrography and the various ocean sciences — marine geology and geophysics particularly — into a closer working relationship. For some years Dominion Hydrographer F.C. Goulding Smith had been pressing for a new survey ship to be based in the Atlantic Region. In Smith's last year in office, 1957 — and in the first year of his successor, Norman Gray — *Baffin* was finally launched. Van Steenburgh sustained the initiative of these men and it was under his influence and direction that the newest and largest ship in the Fisheries and Oceans fleet, CSS *Hudson* was constructed as a combined hydrographic/oceanographic vessel.

Van Steenburgh did not work single-handedly of course. In 1960 he appointed Doctor William Cameron as the department's first oceanographer and together they were largely responsible for the development of the Bedford Institute of Oceanography which opened in 1962 as the first multidisciplinary institute where oceanographers, hydrographers and marine geoscientists worked together. Ross Douglas explains the melding of the sciences this way: "There never was a gulf between hydrography and the other sciences as we might have thought in the old days. Hydrographers are interested in discovering how the ocean floor is shaped; the marine geologist is interested in how it got that shape. The data one gathers is important to the other."

The integration of hydrography and the ocean sciences was exemplified by the appointment in 1967 of Doctor A.E. Collin, a former oceanographer, as Dominion Hydrographer. Cameron's dedication to the concept is further illustrated in the contents of a memorandum he wrote in 1970 to Doctor B.D. Loncarevic, then acting director of the Atlantic Oceanographic Laboratory at BIO. The memorandum, on the subject of integrated specialists, read in part:

> Your proposal that Mr. Ewing should take charge of the Hydrographic/Geophysical project to be carried out this summer in Viscount Melville Sound from a D.O.T. icebreaker [is] most commendable. . . . it represents a

> culmination of a long . . . growth toward . . . true integration of the several disciplines . . . particularly of the development of specialists with broad experience and competence in associated disciplines.
>
> When I first proposed this concept . . . many of my colleagues had serious misgivings as to its propriety or practicability. . . . what success we have already achieved . . . is in great measure due to your efforts. . . .

The memorandum went on to say:

> You recommended that Mr. Melanson should assume the post of Chief Scientist on the Valparaiso to Tahiti leg of Hudson '70. This again is clear evidence that the policy . . . is workable . . . this nomination is a credit to Mr. Melanson who has so effectively won the confidence of his scientific colleagues. . . .

Although Bosco Loncarevic undoubtedly appreciated this accolade, the principle of cooperation between the sciences was not new to him: he and hydrographer Harvey Blandford had carried out the first multidisciplinary survey in 1964 and both were well aware of the benefits to be derived.

Van Steenburgh's approach also injected new enthusiasm into a small but eager group at Ottawa headquarters, a group that had been studying the development of new methods and instruments. About 1963 Adam Kerr, now regional director of hydrography at BIO but then newly shore-based after a stint as combined master/hydrographer-in-charge aboard CSS *Cartier*, was dispatched on a study tour of the United States and Europe. He returned afire with the merits of computerized, automated hydrography.

The original R&D group grew — among others it included at various times Neil Anderson, Mike Bolton, Ross Douglas, Mike Eaton, Hiro Furuya and Reg Gilbert in addition to Kerr. It spun off other development groups in the Atlantic, Central and Pacific Regions. (Quebec region had not then been established.)

From the various R&D groups have come the majority of the managers in today's CHS. And from the groups, also has sprung almost all of the innovative techniques and instrumentation described in earlier chapters. Not the least of these innovations is BIONAV.

This navigation system is centred on a computer that assimilates the input from several positioning systems such as Loran-C and satellite positioning methods. It combines these data with figures from the ship's gyro compass, data from the engineroom, and "number crunches" the whole into a positional fix of stunning accuracy.

With the push from van Steenburgh the Canadian hydrographer's competence changed radically and — within the context of one hundred years of the CHS's existence — almost overnight.

William J. Stewart and Henri Parizeau were both educated men and they were, before anything else, surveyors. Stewart as chief hydrographer and Parizeau as Pacific regional chief each had to deal with the higher levels of bureaucracy and each learned to do it. Each in his time became thoroughly skilled in the use of the new tools as they became available to him. But these were skills that were applied, layer by layer, to the bedrock basis of hydrographic surveying: they did not come naturally.

Today's hydrographers — especially those who occupy the executive offices at Ottawa and the four regional headquarters — *begin* with the skills Stewart and Parizeau worked to attain.

One observer who has witnessed the transformation puts it this way: "Today's senior hydrographer is a technocrat. Oh, sure, out in the field he still slops around in salt-stained jeans and parka. But ashore he wears a business suit and carries a briefcase. He can read balance sheets as easily as he reads a computer printout. He knows the value of working lunches and knows how meetings are run — and how to guide the discussions. Like the equipment he uses and has helped develop, today's hydrographer has become vastly more efficient, very sophisticated, very much the modern technocrat."

In the last twenty to thirty years, change, impelled by technology that advances daily, has swept hydrography out of the course it had followed for two hundred years and more. The men who practise the science have been altered as well, and so far as it is practicable or safe to predict the future, the trend seems likely to continue. In the next and last chapter we will peer into the crystal ball, attempt to blow away the mist and see what hydrography may be — just possibly — in the future.

CSS Hudson

To reach the Western Arctic, the CSS *Hudson* sailed south from its home berth at the Bedford Institute of Oceanography at Dartmouth, Nova Scotia, through the Panama Canal and north to the Institute of Ocean Sciences at Patricia Bay, British Columbia. There she was specially refitted with four hydrographic launches and sophisticated computer technology. The survey set out for the Beaufort Sea where, during the brief Arctic summer, the launches worked 18 hours a day, criss-crossing the ten-mile wide proposed navigation route through the Beaufort. Pingos, or ice cored mounds, litter the Beaufort Sea floor and must be identified to ensure safe navigation will be possible for the projected shipment of oil and gas to southern waters.

While the Beaufort Sea is free of ice for the few weeks of summer, a few hundred miles to the East, the ice is a constant presence. It is this beautiful but hostile environment where much of the hydrographic work will be concentrated during the next decade.

BOLD
BRAVE
BRISK

BOLD

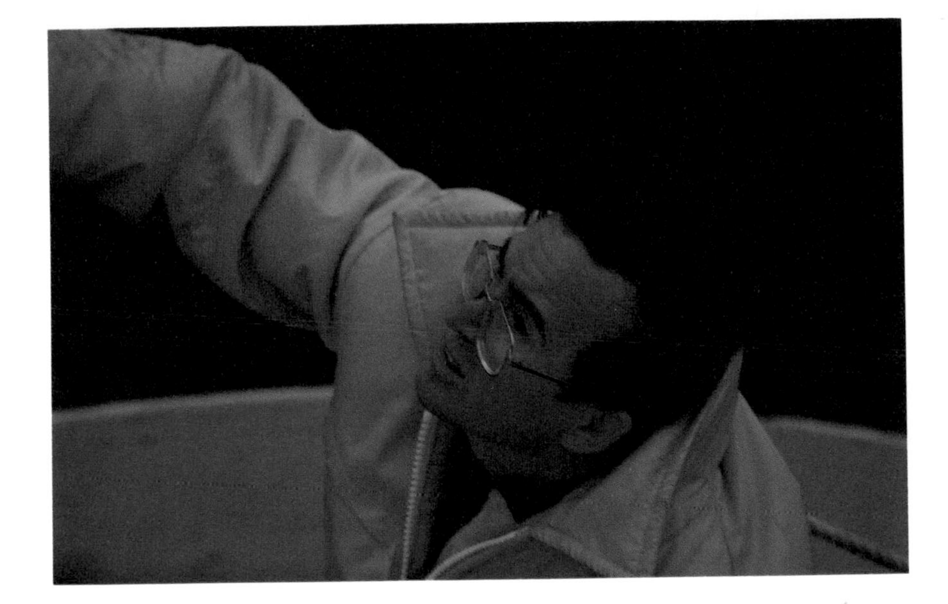

SKIPPER 802
BOTTOM TRACKER

MEMBER
CLUB

Canada
BOLD

HUDSON
OTTAWA

The Promise

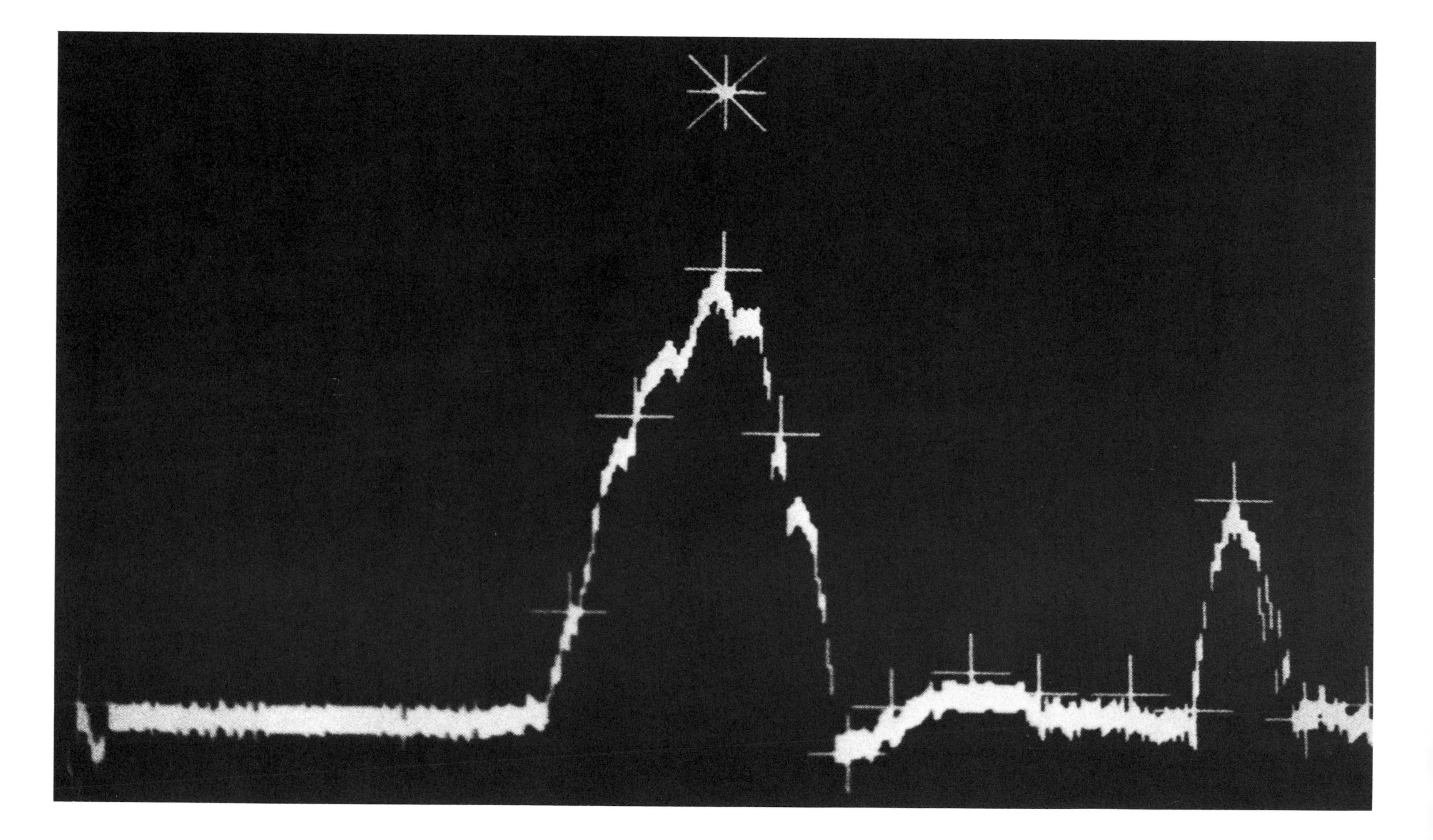

Roll on thou deep and dark blue ocean, roll!
Ten thousand fleets sweep over thee in vain;
Man marks the earth with ruin — his control
Stops with the shore.

Lord Byron (1788-1824)
Childe Harold's Pilgrimage

The oceans do not separate continents but make links between nations.

Igor Mikhaltsev
Deputy Director
Russian Institute of Oceanology

WRITTEN MORE THAN ONE HUNDRED AND FIFTY years ago, Byron's words are sadly ironic now in the face of his own death by drowning and today's pollution which is man's "ruin" of the earth's oceans. Yet they still convey that traditional and enduring sense of humanity's awe of and respect for the vast, seemingly limitless and indomitable, power of the sea. This book has attempted to portray a one-hundred-year history of Canadian hydrographers' attempts to know, to chart and ultimately to domesticate the oceans, lakes, and rivers of this country. We've seen that these men have followed in the wake of predecessors dating back to the earliest mariners of the Orient, the Middle East, and Europe. We've seen how these hydrographers' methods, instruments, charts, ships and the men themselves have changed — sometimes radically — over the past hundred years.

Through all of this history and change, however, at least one thing has remained constant. That is, that, for the greater part, the work of the hydrographer is tied directly to the commercial requirements of whatever power is footing the bill for this work which has always been expensive and time-consuming.

The early charting of Canada's east coast and the Gulf and River Saint Lawrence was done to ensure safe passage of goods and people during the first years of colonization. Then, safe routes were charted for the transport of furs from the interior. Cook's charting of the waters of the Saint Lawrence helped the British conquer Quebec. Cook's discovery and Vancouver's subsequent char-

A video presentation of a Pingo (left).

Data capture and storage system on board a survey launch.

ting of the west coast cleared the way for the fur, lumber and fishing industries in British Columbia. Boulton's Georgian Bay and Great Lakes surveys were essential for the development of shipping on the lakes. The early charting of Canada's Arctic was a result of Britain's search for a shortcut trade route through a much fabled Northwest Passage. The surveys of Ungava in the north were a direct result of the demands of the iron ore mining industry. Today's exploration and charting of the Beaufort Sea have become priorities of the Canadian Hydrographic Service because of the need for Canadian self-sufficiency in oil and gas. And the growing number of requests, since the end of World War II, for charts made especially for the recreational boater is itself a result of the commercial requirements of an expanding tourist/recreation industry.

Always the hydrographer is ultimately the servant of the merchant — and is indentured to the needs of the future. In time of war only, are the demands of commerce pushed aside and the chartmaker concerned merely with present and pressing needs. In general terms, however, work done today must serve the requirements of someone else tomorrow — two years from now, five years, ten, twenty, fifty. And because the pace of economic and technological change has so accelerated in the latter half of this century, the task of deciding which work done today will satisfy the needs of tomorrow becomes at best a guessing game. Do nautical surveyors seek out channels that will accommodate ships with deep drafts of thirty metres, unimaginable tankers drawing up to one hundred metres, or gigantic, million-tonne submarine carriers?

So our hydrographers make educated guesses based on the current status of work already completed and on the requests for future surveys. These latter come from such sources as the mining and petroleum industries, yacht clubs and tourist region authorities, shipping companies, individual taxpayers, and government departments, in particular those of defence and transport.

As of 1982, only forty-five percent of Canada's waterways are surveyed adequately for current requirements. Only fifteen percent of our Arctic waters have been surveyed. And, astonishingly, twenty-five percent of all our waters remain unsurveyed or surveyed only to a minimum, "reconnaissance" level. The main reason that less than half of Canada's shipping lanes have been properly surveyed is the same reason that the CHS can respond to only one third of the legitimate requests for surveys that it receives annually — a scarcity of funds for the overwhelming task of charting the waters of the world's longest national coastline.

Most of what the CHS needs to accomplish in the near future is affected by budgetary restrictions. Beyond those needs of hydrographers in the field, the cartographic programs designed to convert charts to the new format, to make all charts bilingual (at present only twenty-five percent have been translated), to convert to a uniform metric format (approximately twenty percent have been converted), and to create a comprehensive data base for the consolidation of computer-assisted cartography all demand an investment of time and money which is not always forthcoming.

But, historically, chartmakers have never had enough money or manpower. From Columbus, to Bayfield, to the present, the search for sponsors and adequate funding has been and remains a large part of the job of any explorer, of any head of a hydrographic service.

Cognizant of this history and buoyed by the examples of the last fifty years wherein hydrographic advances have been the result, for the most part, of new and more efficient hardware, today's hydrographic service has made the research and development of hydrographic technology a priority in the planning of its own future.

For example, pingos, those underwater mountains of ice discovered in 1969 in the Beaufort Sea, present a serious danger to the deep draft tankers expected to be shipping oil out of the Beaufort by the early 1990s; because the pingos appear to be spread indiscriminately through this sea, it's necessary that one hundred percent of that area be surveyed. To achieve total bottom coverage of such a large body of water and discover the relatively small pinnacle-like features — and to accomplish this end within the short Arctic operating season — has caused the CHS to place what is known as "electronic sweeping" high on its research priority. One method now being investigated is to stream up to four paravane floats, two on either side of the ship, at distances of about 75 and 150 metres. Each float carries its own transducer which transmits and

Snorkel of Dolphin.

. . . a large percentage of the charts of the Western Hemisphere are based on reconnaissance surveys conducted by Great Britain, France, Spain, and the United States, from 50 to 100 years ago. . . . The United States, Canada, Brazil and Argentina have done a great deal during the past forty years to remedy this sad situation, but . . . it will be at least another century before we consider the Americas adequately surveyed from a hydrographic standpoint.

G. Medina
Senior Cartographic Engineer
Hydrographic Office, Washington D.C.
1941

Hydrographer examines a sounding roll.

receives back the bottom information, and passes it to the ship's data logger. Thus one line of ship sounding produces a 300-metre swath of bottom data, also saving time and fuel.

On larger scale surveys a 10-metre launch at BIO carries an extensible boom that projects as much as fifteen metres out from each side of the boat. Light in weight and looking much like a household television antenna mounted horizontally, it carries multiple SONAR transducers that can sweep as much as a one hundred-foot channel.

Recent refinements in the capabilities of what is known as side scan SONAR have proven more useful in the development of the concept of electronic sweeping. Side scan SONAR, first used in England in the 1950s, involves the use of a towed body radiating electronic impulses — not only downward, but also from either side — in a controlled swath pattern over a broad area of the ocean floor. Echoes returning from the sea floor are reflected with varying degrees of intensity dependent upon the angle of incidence between the outgoing signal and the sea floor. The graphic record thus obtained portrays an almost photographic image of the area being surveyed.

In the central Arctic, hydrographers from Burlington use a tethered submersible, lowering it through a hole chopped in the ice. Once lowered through the opening the submersible cruises out from the hole, under the ice, at a predetermined water depth. Constant readings from its SONAR equipment are transmitted back along the cable tether to recording instruments on the ice. The cable may be as long as two kilometres (although CCIW is presently using one only half that length). When the submersible reaches the end of its tether it turns to one side, executes a small arc of a circle, and then returns to the centre hole where it reverses itself and cruises out again on another radius.

Repeating this process the submersible can cover a circular area of the seabed four kilometres in diameter in a fraction of the time it would take the hydrographers to sound the bottom using more conventional through-the-ice techniques. One of the greatest problems with the tethered submersible is keeping the ice hole open in the Arctic's freezing temperatures.

New as it is, the tethered submersible is already somewhat outdated. As the CHS celebrates its one hundredth anniversary it is building an automated remote-controlled submersible. ARCS, as it's known, is a large, torpedo-shaped, remote-controlled submersible that will be lowered through a hole in the ice to sound an area as large as eight kilometres square. It will be propelled by an electronic motor powered by a bank of batteries; it will be steered and controlled by an acoustic telemetry system that will also sample the data. Initially ARCS will be used for sounding only, but it may be adapted later to perform other functions as well. The system is based on a prototype developed at the University of Washington, and it will undergo its first sea trials in the Arctic in 1984.

ARCS sounding under the ice.

The latest of the new systems is SEABED II being developed by the Atlantic Geoscience Centre, at BIO, with some input and funding from the CHS. It makes use of the successful technology developed for shallow seafloor profiling in SEABED I; the new version will combine that technology with side scan SONAR to provide detailed sea floor coverage. The transducers are carried in a "fish" that is towed close to the sea floor. Two models are being developed — one for shallow work to 500 metres, the second for depths to 2,000 metres. CHS's main interest is in the shallow model; the 2,000-metre version is being tested in 1983.

Base camp from which ARCS will be deployed.

The device will also send pressure waves into the sea floor itself. The return "echoes" will give an indication of the seabed's composition — invaluable knowledge in the search for underwater natural resources.

Another area of research and development being given a similarly high priority is that of remote sensing from the air. Particularly exciting is the use of laser beams both to establish the height of an airplane above the water and measure the depth of inshore waters up to twenty metres deep. The sounding of waters close to shore has traditionally taken up most of a field hydrographer's time because of the amount of detailed information that must be gathered for the compilation of charts on the largest of scales. According to Dominion Hydrographer Stephen B. MacPhee, in the near future:

> the task of the hydrographer will change to verifying the bathymetric and foreshore plots produced by the new laser system . . . [And, this new] system has the potential to produce, in a few days' flying time, enough data to keep our field parties going for several seasons.

Other remote sensing experiments have resulted in the deployment and interpretation of colour aerial photography for the establishment of high- and low-water lines and the bathymetry of shoal and shore areas in depths up to seven metres.

Other "futures" projects currently under development include the fine tuning of long-range positioning systems in the Arctic, programs to gather comprehensive tidal data in the far north, the creation of new chart schemes and formats to accommodate better the anticipated needs of the users of CHS charts, and a continued effort to equip all the regional offices with sufficient hardware and personnel to allow for the universal implementation of computer-assisted cartography.

In the long run — probably a *very* long run — the end product of all of this research and development may well be what has been termed the "electronic chart". With all charting data collected and compiled in a digital format and stored in a ship's computer, with information from long-range positioning equipment (onboard receivers for the already established Loran-C network and the soon to be operational, satellite-based Global Positioning System), along with that data provided continuously by a ship-based radar system and the ship's gyrocompass all being fed into the same computer, a video terminal on the bridge will display not only the digitized chart information but also the ship's position and course, and the position of any obstacles or hazards in the immediate vicinity.

The electronic chart may be especially useful in Arctic shipping lanes with their shifting ice conditions and lack of navigational aids, and in harbour navigation where pilots require a navigational capability which can keep track of moving objects and display large scale charts accurately.

Industry confidence in the inevitability of the electronic chart and its ability to revolutionize the ancient art of navigation are such that a 1982 publication, "The Electronic Chart", prepared by the Department of Surveying at the University of New Brunswick, states that:

> The capability of the electronic chart to operate with an interface to the depth sounder and VHF radio could eventually lead to automatic pilots

which determine position with respect to depth contours, and automatically steer the ship by means of a VHF radio link.

We may not see automatic ship's pilots being used universally in this century; however, that same UNB report asserts that some form of electronic chart is just on the horizon. By the year 1990 electronic technology will have advanced to the point where an electronic display of paper chart quality will be feasible for many applications.

Funding for research and development of these systems, for the acquisition and training of more personnel, for ships, and, increasingly, for the contracting out of some survey work to the private sector has been, is and will likely remain limited. Such is the lot of the hydrographer — past, present, and, most probably, future.

How the individual hydrographers of the future will differ from today's chartmakers is a matter for much speculation. In the past we have seen that mariners have been taken aboard and taught surveyors' skills — and vice versa. Over the years, the roles of surveyor and cartographer have become more and more divorced from each other. This may change with the movement of the cartographic function out to the regions and the consequent re-involvement of field personnel in the chartmaking process. The importance of gathering and compiling hydrographic information in digital form has placed and will continue to place greater emphasis on computer-related skills among the ranks of Canadian hydrographers. The trend to the use of multipurpose ships, fulfilling both hydrographic and oceanographic functions, may mean that future hydrographers will be required to possess much of the knowledge and many of the skills of the oceanographer, the pure research scientist.

All of these trends indicate that less and less will the work of a hydrographer be so physically demanding as it has been in the past. This opens the door to the increased participation of women in the hydrographic service. Such a suggestion would no doubt have caused much laughter in the captain's quarters of HMS *Discovery*. It would have been scoffed at in the wardroom of the *Acadia*, and ridiculed even on the decks of the *Baffin* as she made her early voyages. But a hundred years ago who could have envisioned the invention

Processing survey data on board the CSS Hudson.

of the echo sounder, or electronic positioning systems, or the importance of Arctic shipping lanes for oil tankers in the Beaufort Sea?

The hydrographer remains today both a debtor to the past and a servant of the future. Caught between these two, the men and women of the hydrographic service can continue only to do their best. One thing above all else remains constant — a commitment to excellence, a continual striving for nothing less than perfection. The future, a hard taskmaster, awaits and eventually will sit in judgement.

We can only pay our debt to the past by putting the future in debt to ourselves.''

Lord Tweedsmuir
Governor General of Canada, 1935-40

Chief Hydrographers of Canada

Captain John George Boulton, R.N.
Officer-in-charge
Georgian Bay Survey
1883-1893

William James Stewart
Chief Hydrographic Surveyor
Hydrographic Survey of Canada
1893-1925

Captain Frederick Anderson
Dominion Hydrographer
Canadian Hydrographic Service
1925-1936

Frederic H. Peters
Dominion Hydrographer
1936-1947

Robert James Fraser
Dominion Hydrographer
1947-1952

F.C. Goulding Smith
Dominion Hydrographer
1952-1957

Norman G. Gray
Dominion Hydrographer
1957-1967

Arthur E. Collin
Dominion Hydrographer
1967-1972

Gerald N. Ewing
Dominion Hydrographer
1972-1978

Stephen B. MacPhee
Dominion Hydrographer/Director General
1979-

Index

Bibliography

Akrigg, G.P.V. and Akrigg, H.B. *British Columbia/1847-1871: Gold and Colonists.* Vancouver: Discovery Press, 1977.

Appleton, Thomas. *Usque Ad Mare*. Ottawa: Department of Transport, 1968.

Asimov, Isaac. *The Shaping of North America*. London: Dennis Dobson, 1974.

Blewitt, Mary. *Surveys of the Sea*. London: McGibbon and Kee, 1957.

Boulton, J.G. "Hydrographic Surveying". In *Proceedings of the Annual Meeting of the Association of Dominion Land Surveyors*. Ottawa: 1890.

______ . Untitled paper. In "Transactions, 1908-9". Quebec: Literary and Historical Society of Quebec, 1909.

Bowditch, Nathaniel. *Bowditch for Yachtsmen*. New York: David Mckay Company Inc., 1976.

Brown, L.A. *The Story of Maps*. New York: Bonanza Books, 1949.

Burrows, E.H. *Captain Owen of the African Survey*. Rotterdam: A.A. Balkema, 1979.

Canadian Hydrographic Service. *Activity Report*. Ottawa, 1973, 1974, 1975, 1976, 1977, 1978, 1979, 1980.

Casey, M.J. "The Asia Tragedy". *Lighthouse*, no. 12, November 1975.

Dawson, L.S. *Memoirs of Hydrography*. Facsimile reprint, 2 vols. in 1. London: Cornmarket Press, 1969.

Day, Archibald. "Hydrographic Surveys: the purpose and choice of scale". *International Hydrographic Review*, May 1955.

______ . *The Admiralty Hydrographic Service*. London: Her Majesty's Stationery Office, 1967.

Delanglez, Jean. "Franquelin, Mapmaker". *An Historical Review*, vol. 25, no. 1 (new series vol. 14). Chicago: Loyola University, 1954.

______ . *Life and Voyages of Louis Jolliet 1645-1700*. Chicago: Institute of Jesuit History, 1948.

Dunlap, G.D. and Shufeldt, H.H. *Dutton's Navigation and Piloting*. 12th ed. Annapolis, Maryland: Naval Institute Press, 1972.

Dyde, B.S. "From Hand-lead to Hydro-search". In *Proceedings of the 17th Annual Canadian Hydrographic Conference*, April 1978.

Edgell, I. *Sea Surveys: Britain's Contribution to Hydrography*. London: The British Council/ Longmans Green & Co., 1948.

Edmonds, Alan. *Voyage to the Edge of the World*. Toronto: McClelland and Stewart, 1973.

Fagerholm, P.O. "The Parallel Sounding Technique". *International Hydrographic Review*, July 1964.

Friendly, Alfred. *Beaufort of the Admiralty: The Life of Sir Francis Beaufort 1774-1857*. London: Hutchinson, 1977.

Goldsworthy, E.C. "Accurate Control by Electronic Means." *Ship-building and Shipping Record*, 14 September 1967.

Hale, John R. *Age of Exploration*. Great Ages of Man. New York: Time-Life Books, 1966.

Hamilton, Angus C., and Nickerson, Bradford G. "The Electronic Chart". Report, workshop at University of New Brunswick, June 1982.

Hough, Richard. *The Murder of Captain James Cook*. London: Macmillan, 1979.

International Hydrographic Organization. *Hydrographic Dictionary*. 3rd ed. Monaco: 1974.

Jolicoeur, T. and Fraser, J. Keith. "Geographical Features in Canada Named for Surveyors". *Gazetteer of Canada*, Supplement No. 2 Ottawa: Department of Mines and Technical Surveys, Geographical Branch, 1966.

Kemp, Peter, *ed. The Oxford Companion to Ships and the Sea*. London: Granada, 1979.

MacPhee, S.B. *Underwater Acoustics and Sonar and Echo Sounding Instrumentation*. Ottawa: Canadian Hydrographic Service, Technical Report 1, 1979.

______ . "National Hydrographic Surveying and Charting Program — Status and Techniques". In *Proceedings, Third Colloquium of the Canadian Petroleum Association*, Banff, October 1981.

McKenzie, Ruth. "Admiral Bayfield: Pioneer Nautical Surveyor". Ottawa: Environment Canada, Fisheries and Marine Service Miscellaneous Publication 32, 1976.

Meehan, O.M. *The Canadian Hydrographic Service: A Chronology of Its Early History Between the Years 1883 and 1947*. Unpublished manuscript.

Miertsching, Johann. *Frozen Ships*. Translated and introduced by L.H. Neatby, Toronto: Macmillan, 1967.

Mixter, George W. *Primer of Navigation*. Toronto: D. Van Nostrand (Canada), 1967.

Nicholson, N.L., and Sebert, L.M. *The Maps of Canada*. Folkestone (Kent), England: Wm. Dawson & Sons, 1981.

Ollard, Richard. *Pepys*. New York: Holt, Rinehart & Winston, 1974.

O'Shea, J, Champ, C.G., Logan, R.F., MacPhee, S.B. "The Development of Chart Scheming in the Canadian Hydrographic Service". No publishing data supplied.

Peskett, Ken A. "Computer-Assisted Cartographic Station as a Tool for the Cartographer in the Production of Nautical Charts". In *Proceedings of the 19th Annual Canadian Hydrographic Conference*, May 1980.

Peters, F.H. and Smith, F.C. Goulding, "Charting Perils of the Sea". *Canadian Geographical Journal*, February 1946.

Porter, Robert P., "Acoustic Probing of Ocean Dynamics". *Oceanus*, vol. 20, no. 2, Spring 1977.

Pritchard, J.S. Early French Hydrographic Surveys In The St. Lawrence River, *Inter-*

national Hydrographic Review, Monaco LVI (1), January 1979.
Pullen, Hugh F. *The Sea Road to Halifax*. Halifax: Maritime Museum of the Atlantic, 1980.
Ritchie, George Stephen. *Challenger: The Life of a Survey Ship*. London: Hollis & Carter, 1957.
______ . "Great Britain's Contribution to Hydrography During the Nineteenth Century". *Journal of the Institute of Navigation*, vol. 20, no. 1, January 1967.
______ . *The Admiralty Chart: British Naval Hydrography in the Nineteenth Century*. London: Hollis and Carter, 1969.
Roy, Antoine. *Rapport de l'archiviste de la Province de Quebec pour 1943-1944*. Quebec: Secretariat de la Province. 1944.
Russell-Cargill, W.G.A., ed. *Recent Developments in Side Scan Sonar*. Cape Town, University of Cape Town, 1982.
Sandilands, Robert W. "The History of Hydrographic Surveying in British Columbia". *The Canadian Cartographer*, vol. 7, no. 2, December 1970.
______ . "Charting the Beaufort Sea". *Lighthouse*, no. 24, November 1981.
______ . "Captain James Cook, RN, Hydrographer". *The Canadian Surveyor*, December 1978.
______ . "Charlie, Golf, Foxtrot and Quebec". *Lighthouse* Number 20, November 1979.
______ . "Hydrographic Surveying in the Great Lakes during the Nineteenth Century." *The Canadian Surveyor*, June 1982.
Schwartz, Seymour I. and Ralph E. Ehrenberg. *The Mapping of America*. Harry N. Abrams, Incorporated, New York, 1980.
Thomson, Don W. *Men and Meridians*; Volume 1, Department of Mines and Technical Surveys, 1975.
Wilford, John Noble. *The Mapmakers*. New York, Alfred A. Knopf, 1981.

Acknowledgements

It is almost obligatory for authors of non-fictional works to pay tribute to the many persons whose names do not appear within the pages of the finished work but who, nonetheless, have contributed to its contents. A full list of those who have willingly and enthusiastically contributed to this book would strain the limits of this page; to those left unnamed we express our gratitude.

O.M. Meehan of Ottawa is a former hydrographer and for many years served as the unofficial historian of the Canadian Hydrographic Service. His manuscript of the service's early history has been an invaluable source for Chapters 1, 2 and 3.

The Chartmakers was initiated in the mid-1970s by Gerald N. Ewing, the Dominion Hydrographer of the time. Since then Mr.Ewing has moved on to become Assistant Deputy Minister, Ocean Sciences and Surveys, in the Department of Fisheries and Oceans. He has retained his interest in the project and his continued support is appreciated.

In the earliest stages of preparing the text, Bob Langlois assumed responsibility for researching the first four chapters of the book. He submitted his research in a narrative form that became the initial draft for each of these chapters. Later, when the writing was nearly complete, he wrote the first draft of Chapter 6. The authors reworked the initial drafts but a greal deal of Bob Langlois's original work remains. We are deeply and lastingly indebted to him.

The present Director General (Dominion Hydrographer), Stephen B. MacPhee, has maintained the initiative with great vigour. We thank him for his commitment of time, advice and energy. Mr. MacPhee and his four regional directors have read the text and offered suggestions for its improvement. We are indebted to all five. At CHS headquarters in Ottawa Cyril Champ has been an indefatigable searcher-out of elusive facts and his assistance is gratefully acknowledged.

Joanna Drewry of the Communications Branch, DFO, has been the project officer for this publication. We are sincerely grateful to her — and to her colleagues Ian Hamilton and George Sanderson — for their unflagging interest and leadership.

In ways too diverse to enumerate, the following individuals contributed to the work on the book; we thank each: Lieutenant Commander Andrew David RN, Taunton, England; Judge David A. Anderson and Roland G. Andrews, Toronto; the Reverend Father S.J. Pouliot and Patrick Hally, Quebec City; David Avey, Bob Brooks, Bert Smith and Niels Jannasch, Halifax; Bob Young and Fred Smithers, Victoria.

We gratefully acknowledge the permission extended us by Professor Robin S. Harris, Toronto, to quote from the Bayfield-Harris correspondence.

Many individuals have read the manuscript at various stages of completion and made suggestions to improve and correct the text. Any errors of fact or interpretation that remain are the fault of the authors alone.

S.F.
R.W.S.
October 1983

Picture Credits

B.C. Provincial Archives: 35
Canada Post Corp.: 31
Foster, Michael: 1, 41-56, 70, 101-112, 133, 144-159, 178-191, 224-240, 245, 247

Maritime Museum of Nova Scotia: 170-173
National Maritime Museum, Greenwich, England: 16
Parizeau, Mme H.D.: 85
Public Archives of Canada: 5(PA28539), 6(PA28540), 8(C113062), 11(NMC165661), 13(NMC1000-1688), 15(PA112-1702), 21(NMC16854), 59(PA120552)

We would graciously like to thank the Canadian Hydrographic Service for supplying the photographs and charts not credited above.

The text chapters of
The Chartmakers were edited by
Stanley Fillmore.
Peter Maher
of Maher & Murtagh Inc., Toronto,
designed the book using Garamond Book typeface.
Typesetting was by
Nancy Poirier Typesetting Limited, Ottawa,
Alpha Graphics, Toronto and
Jay Tee Graphics Ltd., Toronto.
The book was printed on
Sterling Matte
and bound by Imprimerie gagné ltée,
Louiseville, Quebec.